SpringerBriefs in Computer Science

W0225690

For further volumes:
http://www.springer.com/series/10028

Lu Wang • Kaishun Wu • Mounir Hamdi

Attachment Transmission in Wireless Networks

 Springer

Lu Wang
Department of Computer Science
and Engineering
Hong Kong University of Science
and Technology
Kowloon, Hong Kong SAR

Kaishun Wu
Department of Computer Science
and Engineering
Hong Kong University of Science
and Technology
Kowloon, Hong Kong SAR

Mounir Hamdi
Department of Computer Science
and Engineering
Hong Kong University of Science
and Technology
Kowloon, Hong Kong SAR

ISSN 2191-5768   ISSN 2191-5776 (electronic)
ISBN 978-3-319-04908-3 ISBN 978-3-319-04909-0 (eBook)
DOI 10.1007/978-3-319-04909-0
Springer Cham Heidelberg New York Dordrecht London

Library of Congress Control Number: 2014932668

Printed on acid-free paper

Springer is part of Springer Science+Business Media (www.springer.com)

Preface

Wireless penetration has witnessed explosive growth over the last two decades. Accordingly, wireless devices have become much denser per unit area, resulting in an overcrowded usage of wireless resources. To avoid radio interferences and packet collisions, wireless stations have to exchange control messages to coordinate well. The existing wisdoms of conveying control messages could be classified into three categories: explicit, implicit, or hybrid. However, all these methods consume valuable communication resources, for example, control frames and data packets are transmitted in an alternate manner, either in time domain or in frequency domain, which introduces massive coordination overheads. Therefore, providing cost-effective coordination mechanisms becomes a critical problem in wireless design.

In this book, we present a novel PHY layer technique termed Attachment Transmission, which provides an extra control panel to deliver control messages with minimum overhead. In a traditional transmission paradigm, control messages compete for communication resources with data packets. On the contrary, attachment transmission enables control messages to be transmitted along with data packets, without degrading the effective throughput of the original data packets. In addition to the basic design, this book presents the design challenges, the theoretical model, and demonstrates the implementation on a GNU radio testbed. Extensive experiments demonstrate that attachment transmission is capable of exploiting and utilizing channel redundancy to deliver control messages, thus providing significant support to numerous higher layer applications.

To demonstrate the effectiveness of the attachment transmission, we apply it to a number of classic problems in wireless networks, including the multichannel allocation problem in OFDMA-based networks, the hidden and exposed terminal problems in ad hoc networks, and the multiple access problem in wireless local area networks (WLANs). For the multichannel allocation problem, attachment transmission provides cost-effective identifier signals. These identifiers help mobile stations learn the channel allocation strategy by themselves, and thus achieve cooperation without coordination. For hidden and exposed terminal problems, attachment transmission offers accurate Channel Usage Information (CUI) on who is transmitting or receiving nearby. Therefore, wireless stations can identify hidden and exposed nodes in real time, and thus make the right channel access decisions. For the multiple access problem, wireless clients deliver transmission requests to the access point (AP) through attachments. Since the requests are "attached" on high speed data transmission, control messages will not occupy any resources, such as the communication channel or the transmission air time. In this way we can achieve lightweight control in WLANs. Besides the above scenarios, we believe that attachment transmission can be further exploited and benefit more communication systems.

Kowloon, Hong Kong SAR   Lu Wang
Kowloon, Hong Kong SAR   Kaishun Wu
Kowloon, Hong Kong SAR   Mounir Hamdi
July 2013

Acknowledgments

The authors would like to acknowledge the partial support of the HKUST Grant RPC10EG21, and Guangdong Natural Science Funds for Distinguished Young Scholar (No. S20120011468), Guangzhou Pearl River New Star Technology Training Project (No. 2012J2200081), Guangdong NSF Grant (No. S2012010010427), and China NSFC Grant 61202454.

The authors would also like to thank Prof. Sherman Shen. Without his help this book would not have been possible.

Contents

Acronyms

ACK	Acknowledge
AST	Access Strategy Table
AT	Attachment Transmission
AP	Access Point
BAM	Binary Amplitude Modulation
CAS	Carrier Allocation Scheme
CE	Correlated Equilibrium
CP	Cyclic Prefix
CSMA	Carrier Sense Multiple Access
CSMA/CA	Carrier Sense Multiple Access with Collision Avoidance
CUI	Channel Usage Information
CRF	Current Receiver Field
CRN	Cognitive Radio Network
CSF	Current Sender Field
CTL	Current Transmission List
CVF	Current Victim Field
ET	Exposed Terminal
FFT	Fast Fourier Transform
FSK	Frequency Shift Keying
HT	Hidden Terminal
IC	Interference Cancelation
IFFT	Inverse Fast Fourier Transform
MAC	Media Access Control
MAI	Multiple Access Interference
MCM	Multi-Carrier Modulation
MS	Mobile Station
MTU	Maximal Transmit Unit
NE	Nash Equilibrium
NHL	Neighborhood Hash List
OFDM	Orthogonal Frequency-Division Multiplexing
OFDMA	Orthogonal Frequency-Division Multiple Access

PRR	Packet Reception Rate
Qos	Quality of Service
RTS/CTS	Request To Send/Clear To Send
SLA	Service Level Agreement
WLAN	Wireless Local Area Network
WMAN	Wireless Metropolitan Area network

Chapter 1
Introduction

1.1 Introduction

Nowadays wireless applications and services have been a close and inseparable part of people's daily lives. As a consequence, wireless devices are being deployed much more densely than before, resulting in a crowded usage of wireless resources. To address this problem, extensive research has been conducted to increase the usage efficiency of wireless resources to mitigate any collisions due to crowded usage of wireless resources. In particular, it has been shown that it is beneficial to divide the current frequency band into many smaller subchannels and let more than one user share a given frequency band. As a result, the whole channel can sustain multiple transmissions at the same time. Orthogonal Frequency Division Multiplexing (OFDM) has been demonstrated to be a promising technique for this paradigm, which is capable of combating inter-symbol interference and providing multiuser diversity gain. However, when dividing the whole channel into multiple subchannels, the allocation of subchannels to different nodes remains first challenge. This is even more critical in distributed networks, which is known as the multichannel allocation problem. In addition, in each fine-grained subchannel, resources may not be fully utilized due to a number of classic problems, such as the multiple access problem and the hidden/exposed terminal problems. In an infrastructure-based wireless network, multiple clients may want to transmit to the Access Point (AP). Therefore, it is critical for the AP to schedule the transmission through massive coordination. For the hidden terminal problem, owing to the inability to figure out whether a current receiver is busy or not in a hidden terminal topology, hidden nodes keep transmitting to a current receiver and thus frequent collisions occur. For the exposed terminal problem, since the exposed nodes cannot distinguish whether their transmissions will interfere with current neighboring transmissions in an exposed terminal topology, they give up their transmissions, although other current receivers are not influenced even when these exposed nodes conduct concurrent transmissions. All these problems call for efficient solutions to increase the efficient usage of wireless resources.

L. Wang et al., *Attachment Transmission in Wireless Networks*,
SpringerBriefs in Computer Science, DOI 10.1007/978-3-319-04909-0_1,
© The Author(s) 2014

In centralized networks, channel allocation is designated by certain authorities, such as AP, and can achieve good utilization of the available channel capacity without incurring much overhead. However, in distributed networks, since there are no such authorities, channel allocation simply relies on extensive coordination among nodes (cooperative) or historical knowledge of themselves (noncooperative). Recently, a lot of research has focused on developing distributed multichannel allocation protocols based on Game Theory. The aim of such schemes is to achieve Nash Equilibrium (NE) of a multichannel allocation game, that is, to achieve the best payoff for all nodes. Mähönen et al. in [1] propose a simple noncooperative scheme for multichannel allocation based on Minority Game, where nodes maintain access strategy for each channel based on transmission history. However, with limited information of other nodes' strategies, their approach does not have the desired NE, where fairness among nodes cannot be ensured. Gao and Wang [2] formalize a multichannel allocation in multi-hop networks as a Cooperative Game. They have so far achieved a good NE, yet have to consume certain resources for coordination.

When trying to solve the hidden and exposed terminal problems, a trade-off arises between collisions (hidden nodes) and unused capacities (exposed nodes). Carrier Sense Multiple Access (CSMA) is the best solution for this trade-off and designs a handshake mechanism called RTS/CTS to solve both the hidden and exposed terminal problems. However, this handshake leads to considerable overhead and brings in new problems when being applied to multi-hop networks, such as false blocking and masked nodes. Also, the information CSMA provides (whether the channel is busy or not) is too coarse. Nodes can barely make the right channel access under a hidden terminal or an exposed terminal topology. Other research efforts have only focused on one particular problem. Full duplex [3] allows a receiver to send a busy tone when receiving a data packet. This scheme mitigates the hidden terminal problem, but the exposed node still exists. CMAP [4] deduces the exposed node and excludes collided transmissions by consulting a "Conflict Map," but the increase in hidden terminal problem is severe. Carrier Sense Multiple Access with Collision Avoidance (CSMA/CA) designs a handshake mechanism called RTS/CTS [5] to mitigate both hidden and exposed terminal problems. However, RTS/CTS is rather costly and introduces other problems such as false blocking. Therefore, RTS/CTS is disabled by default in wireless local area networks (WLANs) and AP manufactures.

The difficulty in solving the above-mentioned problems stems from the trade-off between coordination and control overhead. Since the control messages and data traffic often interleave in wireless communications, it is desired to propose an efficient communication system that serves them together at the same time. As shown in Fig. 1.1, the data traffic and the control messages are transmitted simultaneously in the same channel. Data traffic accounts for the entire transmission air time and is allocated the same bandwidth as in traditional systems. In the meantime, control messages are transmitted in an attached manner with the data traffic, dramatically reducing the coordination overhead.

Recently, Interference Cancelation (IC) techniques [6, 7] have made significant progress to recover transmission errors caused by interference. This gives us an insight to propose a new coding scheme, Attachment Coding, to provide additional

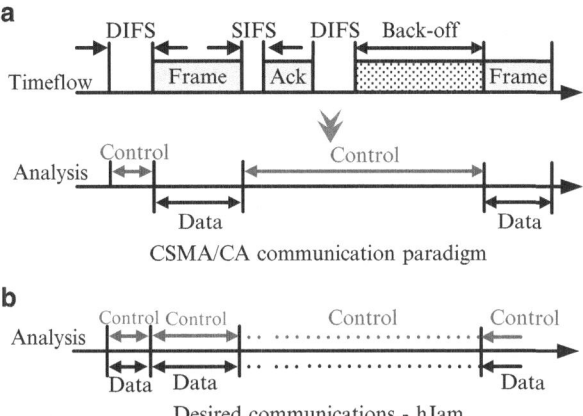

Fig. 1.1 (**a**) An example of a CSMA/CA communication paradigm and a simple analysis of its performance. (**b**) Desired communication system with control messages and data packets being transmitted together

information we require without degrading the effective throughput in the original transmission. Specifically, control information is modulated into specially designed signals called *Attachment*. By intentionally injecting an attachment on one's data packet or transmitting standalone on air, others are capable of acquiring control information through the attachments. This attached manner is promising due to its ability to avoid additional bandwidth for control information.

By successfully applying Interference Cancelation technique in OFDM-based wireless networks, we manage to achieve the *Attachment Transmission* communication system with *Attachment Coding* paradigm. The attached control information can be detected and decoded by anyone who needs it, and it can also be canceled for data recovery. One possible question might be that whether we are always able to find such redundancy for attachment transmission. Since the state-of-the-art rate adaptation algorithm cannot fully unitize the whole channel bandwidth, we argue that data packets are always capable of carrying small amounts of attachment transmissions.

To demonstrate the effectiveness of attachment transmission, we propose several new cross-layer designs to solve the above-mentioned problems, which include:

- *Harmless Attachment* that uses attachment transmission to benefit multiple access in the infrastructure-based WLAN [8]. Clients will "attach" their request on the ongoing data transmissions to the AP, and thus dramatically reduce the contention overhead.
- *Attachment Learning* that helps mobile stations to learn allocation strategies by themselves. After the learning stage, mobile stations can achieve a TDMA-like performance, where stations know exactly when and on which channel to transmit without further collisions.

- *Attachment Sense* that identifies hidden and exposed terminals in ad hoc networks. The self-attached control provides accurate channel usage information in real time, thus it guides mobile stations to make the right channel access decision quickly and accurately.

We verify the feasibility of attachment transmission using a GNU Radio testbed. The experimental results shows that attachment transmission is feasible for cost-effective control information. We also evaluate the performance of Harmless Attachment, Attachment Learning, and Attachment Sense using NS3, python, and C++. Harmless Attachment exhibits significant higher efficiency than prior protocols that do not allow concurrent transmission of coordination and data. This provides up to 200 % gain in efficiency, as compared to traditional 802.11 standards. By utilizing attachment transmission, Attachment Learning can achieve CE of a multichannel allocation game with guaranteed fairness among stations. The average throughput can be improved by up to 300 % over Slotted ALOHA. Meanwhile, Attachment Sense improves the averaged throughput by as much as 200 % over CSMA in ad-hoc networks, verifying that it successfully identifies and resolves both the hidden and exposed terminal problems. Therefore, interference introduced by hidden nodes can be reduced, and more concurrent transmissions that have not been carried out before due to exposed nodes can now be leveraged.

1.2   Outline

This book focuses on the design of attachment transmission and its applications in wireless communication. It also aims to introduce the basic concept about attachment transmission and present some application scenarios in this area. The following is an overview of the book.

Chapter 2: The purpose of this chapter is to provide the necessary building blocks for the design of the attachment transmission and its applications. Included in this chapter are the OFDM modulation primer, reviews of some novel PHY techniques for wireless communications, and related works of some classic problems in wireless networks. As a promising technique, OFDM modulation transmits all the orthogonal carriers simultaneously, and thus is capable of achieving high data rate and combating multipath fading. Since then, numerous PHY layer techniques built on OFDM modulation have emerged to assist MAC layer protocols in recent years. Their main object is to reduce the coordination overhead. These techniques are also designed to solve some classic problems, such as multichannel allocation, hidden terminal and exposed terminal problems. At the end of this chapter, we also present some survey on these problems.

Chapter 3: In this chapter, we describe the overall architecture of an *Attachment Transmission* enabled communication system. Attachment transmission is built on top of an OFDM-based system. It modulates control information into narrow-band signals and transmits them into air without any impact on the original data packets. The design of attachment transmission includes two components: (1) attachment modulation and demodulation and (2) attachment cancelation and data recovery. In this chapter, we first describe the system overview and raise some design challenges. After that, the detailed design of attachment transmission is presented along with design principles in a theoretical way. Finally, we implement attachment transmission on software defined radios (SDRs) to verify the feasibility and reliability.

Chapter 4: In this chapter, we present several typical applications that leverage attachment transmission to solve the above-mentioned classical problems. With the assistance of attachment transmission, a variety of MAC layer protocols are designed, which guide mobile stations to make better access decisions through the PHY layer information they need. These applications include: *Harmless Attachment* that uses attachment transmission to benefit multiple access in WLANs in the infrastructure mode [8]. *Attachment Learning* that helps mobile stations to learn allocation strategies by themselves. After the learning stage, mobile stations can achieve a TDMA-like performance, where stations can know exactly when and on which channel to transmit without further collisions. *Attachment Sense* that identifies hidden and exposed terminals in ad hoc networks. The self-attached control provides accurate channel usage information in real time, thus guiding mobile stations to make the right channel access decision quickly and accurately. In the following sections, we present how we utilize attachment transmission to solve the above-mentioned problems.

Chapter 5: In the previous chapter, we presented the attachment transmission design and demonstrated several promising applications to solve certain classical problems. In this chapter, we first draw up the conclusion of the book and then further discuss potential scenarios that may benefit from attachment transmission, including Quality of Service (QoS) control [9, 10] in multimedia transmission and sensing and coordination in Cognitive Radio network (CRN) [11, 12]. These topics are also worth studying, and we believe that more communication systems could benefit from the attachment transmission paradigm.

References

1. P. Mähönen and M. Petrova, "Minority game for cognitive radios: Cooperating without cooperation," *Physical Communication*, vol. 1, no. 2, pp. 94–102, 2008.
2. L. Gao and X. Wang, "A game approach for multi-channel allocation in multi-hop wireless networks," in *Proceedings of the 9th ACM international symposium on Mobile ad hoc networking and computing*, pp. 303–312, ACM, 2008.
3. M. Jain, J. Choi, T. Kim, D. Bharadia, S. Seth, K. Srinivasan, P. Levis, S. Katti, and P. Sinha, "Practical, real-time, full duplex wireless," in *ACM MobiCom*, pp. 301–312, 2011.
4. M. Vutukuru, K. Jamieson, and H. Balakrishnan, "Harnessing exposed terminals in wireless networks," in *Proceedings of the 5th USENIX Symposium on Networked Systems Design and Implementation*, pp. 59–72, 2008.
5. I. W. Group *et al.*, *IEEE 802.11n-2009: Enhancements for Higher Throughput*, 2009.

6. D. Halperin, T. Anderson, and D. Wetherall, "Taking the sting out of carrier sense: interference cancellation for wireless lans," in *Proceedings of the 14th ACM international conference on Mobile computing and networking*, pp. 339–350, ACM, 2008.

7. S. Katti, S. Gollakota, and D. Katabi, "Embracing wireless interference: Analog network coding," in *ACM SIGCOMM Computer Communication Review*, vol. 37, pp. 397–408, 2007.

8. Y. Bejerano, H.-G. Choi, S.-J. Han, and T. Nandagopal, "Performance tuning of infrastructure-mode wireless lans," in *Modeling and Optimization in Mobile, Ad Hoc and Wireless Networks (WiOpt), 2010 Proceedings of the 8th International Symposium on*, pp. 60–69, IEEE, 2010.

9. C. Zhu and M. S. Corson, "Qos routing for mobile ad hoc networks," in *INFOCOM 2002. Twenty-First Annual Joint Conference of the IEEE Computer and Communications Societies. Proceedings. IEEE*, vol. 2, pp. 958–967, IEEE, 2002.

10. Y. Wu and D. H. Tsang, "Distributed power allocation algorithm for spectrum sharing cognitive radio networks with qos guarantee," in *INFOCOM 2009, IEEE*, pp. 981–989, IEEE, 2009.

11. C. Gao, Y. Shi, Y. Hou, H. Sherali, and H. Zhou, "Multicast communications in multi-hop cognitive radio networks," *IEEE Journal on Selected Areas in Communications*, vol. 29, pp. 784–793, 2011.

12. I. F. Akyildiz, B. F. Lo, and R. Balakrishnan, "Cooperative spectrum sensing in cognitive radio networks: A survey," *Physical Communication*, vol. 4, no. 1, pp. 40–62, 2011.

Chapter 2
Recent Advances in Wireless Communications

In this chapter, we provide some necessary building blocks for the attachment transmission design and its applications, including the OFDM/OFDMA modulation primer and reviews of some novel PHY techniques for wireless communications. Then, at the end of this chapter, we also present a survey on some classic wireless network problems.

2.1 OFDM/OFDMA Preliminary

We first introduce the basic idea of an OFDM/OFDMA-based system. OFDM modulation has been developed into a promising technique for multi-carrier transmissions, which improves the network performance significantly for future wireless communications. OFDM transforms a frequency-selective wide-band channel into a group of nonselective narrow-band channels named subcarriers, which makes it robust against large delay spreads and cross-talk effect by preserving orthogonality in the frequency domain. Orthogonal Frequency-Division Multiple Access (OFDMA) is a straightforward extension of OFDM into a multiuser environment. It has a series of attractive features, including scalability, intrinsic protection against multiple access interference (MAI), as well as flexible resource management. Therefore, it is adopted in a wide range of systems, such as multiuser satellite communications [1] and fourth-generation cellular networks [2].

2.1.1 OFDM Basis

OFDM can be considered as a combination of multi-carrier modulation (MCM) and frequency shift keying (FSK) modulation [3]. MCM is an approach to data transmission that involves dividing the transmitting data stream into several parallel

L. Wang et al., *Attachment Transmission in Wireless Networks*,
SpringerBriefs in Computer Science, DOI 10.1007/978-3-319-04909-0_2,
© The Author(s) 2014

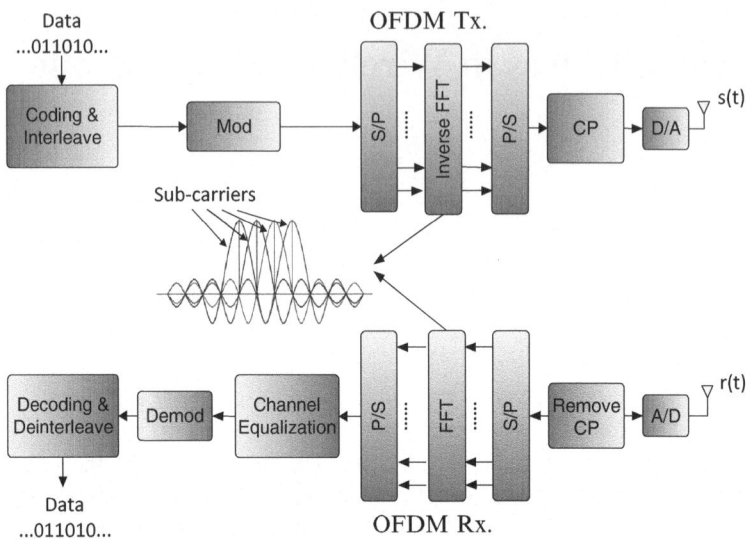

Fig. 2.1 OFDM modulation block diagram. The transmitted signal is spread and modulated across the subcarriers

bit streams. Each bit stream has a much lower data rate, and are modulated onto individual carriers or subcarriers. FSK modulation is a technique whereby data is transmitted through discrete frequency changes of a carrier wave. To achieve orthogonality amongst the carriers, it separates them by an integer multiples of the inverse of symbol duration of the parallel bit streams. When combining MCM and FSK together in OFDM, the entire allocated channel is occupied through the aggregated sum of the narrow orthogonal subbands.

We now detail the transmission procedure. On the transmitter side, the data to be transmitted on an OFDM signal is spread across the carriers of the signal, each carrier taking part of the payload. This baseband modulation is performed via an inverse Fast Fourier Transform (IFFT). To combat symbol misalignment due to multipath effects, OFDM has a built-in robustness mechanism called Cyclic Prefix (CP). Instead of using an empty guard space, a cyclic extension of the OFDM symbol fills the gap, which has a length that exceeds the maximum delay of the multipath propagation channel. After that, the signal sequence with CP is converted into analogue signals and then transmitted into air.

Upon receiving the signals, the receivers sample them and pass them to a demodulation process chain. After the sampling procedure, the sampled data blocks are processed by Fast Fourier Transform (FFT) process and the final result is the original data subject to certain scaling and phase rotations. These scaling and phase rotations are mainly due to channel dispersion. Therefore, channel equalization is needed to recover the original data from the distorted one. Figure 2.1 illustrates the basic structure of an OFDM communication system.

2.1.2 OFDMA Basis

OFDMA is a multiuser version of OFDM, which allows simultaneous low data rate transmissions from several Mobile Stations (MSs or clients). In 2002, OFDMA was adopted as the air interface for emerging IEEE 802.16e standards for Wireless Metropolitan Area Network (WMAN) [4]. In the OFDMA subcarrier structure, the subcarrier frequency spacing is fixed. To support a wide range of bandwidths, it simply adjusts the FFT size to the channel bandwidth. Therefore, the basic unit of physical resource is fixed and the impact on higher layers is minimized. This significantly improves the deployment scalability and flexibility and is one of the most essential features offered by OFDMA.

In an OFDMA-based network, the available subcarriers are divided into several mutually exclusive groups represented by subbands. Each group of subcarriers is assigned to one MS for concurrent transmission [5]. Signals from clients are separated in time and/or frequency domains. That is, the orthogonality among subcarriers ensures that clients are protected from MAI. In particular, time is partitioned into fixed length frames across all the subcarriers. A frame can be of length 2, 2.5, 4, 5, 8, 10, 12.5, or 20 ms [6]. This value depends on several factors, such as channel conditions and the duration of control information. The allocation for each client is performed in the unit of time × frequency, which is called a slot. Hence, multiple clients are allocated different slots in the time and frequency domain, that is, different groups of subcarriers and/or OFDM symbols are used to transmit the signals to/from multiple users. Each MS can have its own expected bit-rate and Service Level Agreement (SLA).

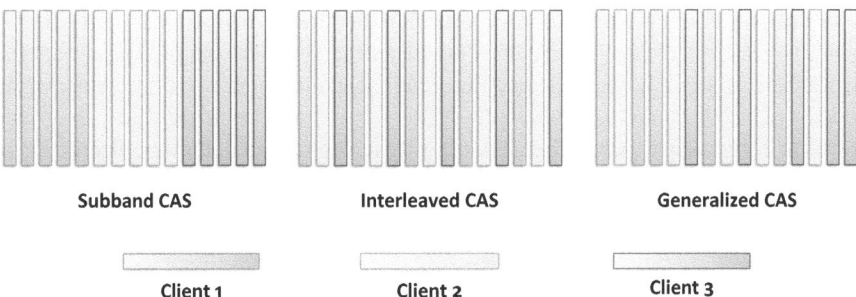

Subband CAS	Interleaved CAS	Generalized CAS
Client 1	Client 2	Client 3

Fig. 2.2 Illustration of three commonly used subcarrier allocation schemes: subband CAS; interleaved CAS; generalized CAS

There are three commonly used carrier allocation schemes (CAS): subband CAS (SCAS); interleaved CAS (ICAS); and generalized CAS (GCAS). We illustrated these three allocation schemes in Fig. 2.2. SCAS means that all the subcarriers of each client are grouped together. In this way, receivers can easily separate the received signals by filter banks. The major disadvantage of SCAS is the inability to leverage frequency diversity. Therefore, the performance of a client could easily be affected by deep fading in some subcarriers. To solve this problem, the ICAS is proposed, which allocates the subcarriers with uniform spacings for different clients. However, since different clients experience different channel conditions and fading, there still exist certain limitations. To overcome the above limitations, GCAS, as the most flexible and desirable scheme, has emerged to meet clients' diverse subcarrier requirements. This allows users to select the best available subcarriers to transmit their packets, enabling the frequency diversity to be fully utilized. However, how to coordinate and negotiate between clients remains a great concern. Therefore, efficient coordination mechanisms are required to take advantage of generalized subcarrier allocation schemes and to achieve maximum throughput and minimize interferences.

2.2   PHY-Layer Assist Communication Paradigms

PHY layer techniques have been frequently used to assist MAC layer protocols in recent years. In [7], a PHY layer RTS/CTS is proposed for multi-round leader election. A PHY layer interference model is proposed in [8] for link scheduling. In [9], the author utilizes a PHY layer ACK to reduce the traditional link layer ACK overhead. Attachment Coding similarly shares the idea of PHY signaling, but differs from the above approaches in that it enables PHY layer control messages to be transmitted simultaneously with data traffic. Therefore, PHY layer control messages do not occupy the bandwidth of the original data traffic, and thus significantly reduces the control overhead. Side channel in [10] uses "interference pattern" for users to jam control information on other's data packets without IC, while FAST simply transmits control information on air and recovers the original data packets from the row signals, which is much more reliable and flexible. Our previous work *h*jam [11] adds jamming signals on other users' packets, in this way they can provide access requests for a certain authority in centralized networks. Therefore, it cannot be used in decentralized networks. In FAST, however, control information is simply transmitted in *Attachments*, which is independent to ongoing data packets. Therefore, it can provide flexible PHY layer information for higher layer protocols and is more suitable for distributed and unsynchronized networks.

2.3   Review of Classic Problems in Wireless Networks

At the end of this chapter, we present some classic problems in wireless networks to better understand the background of attachment transmission. These classic problems include the coordination approaches for wireless communication, multichannel allocation problem, as well as the hidden and exposed terminal problems.

2.3.1 Coordination Approaches for Wireless Communications

In order to address radio interference issues and reduce transmission collisions, a large amount of coordination schemes have been proposed. The existing approaches can be classified as out-of-band and in-band.

The out-of-band coordination approaches are more suitable for multichannel/radio environments. In these approaches, they often allocate a dedicated PHY channel to control messages [12, 13]. Stations switch in and off the control channel during transmissions, leading to significant switching overhead. In addition, these approaches consume an entire channel for control purpose only, which is too expensive. Recently, in [14] the channel contention is moved from the time domain to the frequency domain but requires extra antennas for listening purposes only.

The in-band approaches deliver the control traffic in the same channel as the data traffic. These approaches also consume the communication resources. In the current 802.11 legacy protocol design [15, 16], the coordination is scheduled along the temporal space, which introduces large overheads such as the DIFS, SIFS, and random back-offs. Some recent works also reveal the need for optimal CSMA by experimental results [17]. In [18], they propose a minimum controlled coordination by reducing the DCF overhead. However, in our hJam, we remove such coordination overhead. Also different from CDMA [19] using PN code in code space, our hJam exploits the opportunity in the frequency domain.

2.3.2 Multichannel Allocation Problem

Researchers have long been exploiting multichannel capacity in wireless networks. A lot of wireless standards support multiple channels for concurrent transmissions, such as WiMAX, sensor networks [20], and cognitive radio networks [21]. Traditional methods for multichannel allocation can be classified into four categories: dedicated control channel (e.g., DCA [22] and DPC [23]), split phase (e.g., MMAC [24] and MAP, as presented in [25]), common hopping (e.g., CHMA [26] and CHAT [27], and multiple rendezvous (e.g., SSCH [28]).

Dynamic Channel Assignment (DCA) in [22] is a representation of a dedicated control channel. The overall bandwidth is divided into one common control channel and n data channels. Each node is equipped with a second radio as a control radio, which will operate on the common control channel to exchange control information and to obtain rights to access the data channels. However, under a high traffic load, the control channel becomes a bottleneck. Common hopping is a sophisticated approach to solve the channel reservation problem and improve the channel utilization with single radio, such as Channel Hoping Multiple Accesses (CHMA) [26]. In CHMA, time is divided into small slots, each corresponding to one of the channels. All nodes hop together following a predefined pattern and negotiate their transmissions using the same channel. Whenever a sender/receiver pair agrees to transmit, they will stay in that channel for the rest of the period, while others keep hopping to the next channel. Common hopping has improved channel utilization when compared with dedicated control channel. However, precise synchronization is required among nodes, and switching time for channel hopping also brings considerable cost.

Previous solutions have the common drawback that all nodes waiting to transmit converge on the same channel. With only single rendezvous, the rendezvous channel can become a bottleneck when data packets are not much longer than control packets, or when a large number of channels are available. Thus a multiple rendezvous approach is proposed to overcome this bottleneck, such as SSCH [28]. Although time is still divided into slots as in common hopping, nodes maintain their own hopping patterns and wait for their intended receivers for transmission. This kind of approach effectively mitigates the congestion on the common control channel and is actually the rudiment of game theoretical approaches.

Adaptive subchannel allocation in [29] is the first work to treat resource allocation as an optimization problem for OFDMA. Since then, a considerable amount of research based on Game Theory has been conducted for channel allocation problems. The aim of game theoretical approaches is to balance users' interests, and thus the whole system performance can be improved. This class of approaches can eliminate common control channels in the above-mentioned research, and the coordination overhead can also be significantly reduced. The allocation protocol proposed by Mähönen and Petrova in [30] depends merely on transmission history, thus it greatly reduces the coordination overhead. However, short-term transmission history is not a very good interpreter to adapt channel access. Therefore, it only can achieve a throughput better than Multichannel ALOHA. Park and Van Der Schaar in [31] prove that with enough memory to store transmission history, users can achieve TDMA like performance. However, such memory requirement is too crucial for mobile stations. To achieve efficient multichannel allocation without coordination, Cigler and Faltings in [32] propose a multi-agent learning mechanism for distributed users, where a global coordination signal is predefined for learning. They do achieve Correlated Equilibrium for resource allocation games. However, the coordination signal cannot be easily obtained, and hence the sender/receiver negotiation is not considered in their work.

2.3.3 Hidden and Exposed Terminal Problems

There has been a considerable amount of research on hidden and exposed terminal problems in wireless networks, since these two problems significantly degrade the network performance. A common approach to solve both these problems is to use an RTS/CTS handshake [33], which is also known as "virtual carrier sensing." RTS/CTS handshake utilizes RTS/CTS exchanges to avoid collision in the case of a hidden terminal problem, and infer the transmission concurrency in the case of an exposed terminal problem. Extensive mechanisms then emerge based on the RTS/CTS handshake. MACA-P [34] enhances the RTS/CTS mechanism to increase transmission concurrency. It designs a control gap to synchronize RTS/CTS exchange between different node-pairs, while RTSS/CTSS [35] adds an off-line training phase before RTS/CTS exchanges to further explore transmission concurrency. However, the above RTS/CTS handshake-based mechanisms are not feasible in practice, since RTS/CTS handshake leads to a considerable overhead. Recent work named CMAP [36] proposes an online "conflict Map" to deduce exposed nodes. A special header/tailer is designed for receivers to figure out interferers, and thus allows exposed nodes to transmit concurrently. However, the hidden terminal problem still exists. Full duplex [37] proposes a practical busy-tune scheme to solve the hidden terminal problem, but the exposed terminal problem becomes more severe. Unlike the above approaches, FAST utilizes a PHY layer technique to provide useful Channel Usage Information for higher layers. Therefore, it can solve both the hidden and exposed terminal problems in a cost-efficient way.

References

1. L. Wei and C. Schlegel, "Synchronization requirements for multi-user ofdm on satellite mobile and two-path rayleigh fading channels," *Communications, IEEE Transactions on*, vol. 43, no. 234, pp. 887–895, 1995.
2. M. Tanno, Y. Kishiyama, N. Miki, K. Higuchi, and M. Sawahashi, "Evolved utra-physical layer overview," in *Signal Processing Advances in Wireless Communications, 2007. SPAWC 2007. IEEE 8th Workshop on*, pp. 1–8, IEEE, 2007.
3. J. Heiskala and J. Terry, "Ofdm wireless lans: a theoretical and practical guide," 2002.
4. C. Eklund, R. B. Marks, K. L. Stanwood, and S. Wang, "Ieee standard 802.16: a technical overview of the wirelessman/sup tm/air interface for broadband wireless access," *Communications Magazine, IEEE*, vol. 40, no. 6, pp. 98–107, 2002.
5. M. Morelli, C.-C. Kuo, and M.-O. Pun, "Synchronization techniques for orthogonal frequency division multiple access (ofdma): A tutorial review," *Proceedings of the IEEE*, vol. 95, no. 7, pp. 1394–1427, 2007.
6. I. W. Group *et al.*, *IEEE Standard for Local and Metropolitan Area Networks, Part 16: Air Interface for Fixed Broadband Wireless Access Systems*, 2004.
7. B. Roman, F. Stajano, I. Wassell, and D. Cottingham, "Multi-carrier burst contention (mcbc): Scalable medium access control for wireless networks," in *IEEE Wireless Communications and Networking Conference (WCNC)*, pp. 1667–1672, 2008.
8. P. Wan, O. Frieder, X. Jia, F. Yao, X. Xu, and S. Tang, "Wireless link scheduling under physical interference model," in *IEEE INFOCOM*, pp. 838–845, 2011.
9. A. Dutta, D. Saha, D. Grunwald, and D. Sicker, "Smack: a smart acknowledgment scheme for broadcast messages in wireless networks," in *ACM SIGCOMM Computer Communication Review*, vol. 39, pp. 15–26, 2009.
10. K. Wu, H. Tan, Y. Liu, J. Zhang, Q. Zhang, and L. Ni, "Side channel: bits over interference," in *ACM MobiCom*, pp. 13–24, 2010.
11. K. Wu, H. Li, L. Wang, Y. Yi, Y. Liu, Q. Zhang, and L. Ni, "hjam: Attachment transmission in wlans," in *INFOCOM, 2012 Proceedings IEEE*, pp. 1449–1457, IEEE, 2012.
12. G. Zhou, C. Huang, T. Yan, T. He, J. A. Stankovic, and T. F. Abdelzaher, "Mmsn: Multi-frequency media access control for wireless sensor networks," in *IEEE Infocom*, pp. 1–13, 2006.

13. J. Zhao, H. Zheng, and G.-H. Yang, "Distributed coordination in dynamic spectrum allocation networks," in *New Frontiers in Dynamic Spectrum Access Networks, 2005. DySPAN 2005. 2005 First IEEE International Symposium on*, pp. 259–268, IEEE, 2005.
14. S. Sen, R. Choudhury, and S. Nelakuditi, "Listen (on the frequency domain) before you talk," in *ACM SIGCOMM Workshop on Hot Topics in Networks*, p. 16, 2010.
15. I. W. Group et al., *IEEE 802.11n-2009: Enhancements for Higher Throughput*, 2009.
16. Y. Cheng, H. Li, P. Wan, and X. Wang, "Capacity region of a wireless mesh backhaul network over the csma/ca mac," in *INFOCOM, 2010 Proceedings IEEE*, pp. 1–5, IEEE, 2010.
17. B. Nardelli, J. Lee, K. Lee, Y. Yi, S. Chong, E. W. Knightly, and M. Chiang, "Experimental evaluation of optimal csma," in *INFOCOM, 2011 Proceedings IEEE*, pp. 1188–1196, IEEE, 2011.
18. Z. Zeng, Y. Gao, K. Tan, and P. Kumar, "Chain: Introducing minimum controlled coordination into random access mac," in *INFOCOM, 2011 Proceedings IEEE*, pp. 2669–2677, IEEE, 2011.
19. J. Proakis, *Digital Communications*. McGraw-Hill series in electrical and computer engineering, McGraw-Hill, 2001.
20. Y. Yang, Y. Liu, and L. Ni, "Level the buffer wall: Fair channel assignment in wireless sensor networks," *Computer Communications*, vol. 33, no. 12, pp. 1370–1379, 2010.
21. X. Feng, J. Zhang, and Q. Zhang, "Database-assisted multi-ap network on tv white spaces: Architecture, spectrum allocation and ap discovery," in *IEEE DySPAN*, pp. 265–276, 2011.
22. S. Wu, Y. Tseng, C. Lin, and J. Sheu, "A multi-channel mac protocol with power control for multi-hop mobile ad hoc networks," *The Computer Journal*, vol. 45, no. 1, pp. 101–110, 2002.
23. A. T. Garcia-Luna-Aceves and J. J. G. luna aceves, "Channel hopping multiple access with packet trains for ad hoc networks," in *In IEEE Mobile Multimedia Communications (MoMuC)*, 2000.
24. J. So and N. Vaidya, "Multi-channel mac for ad hoc networks: handling multi-channel hidden terminals using a single transceiver," in *ACM MobiHoc*, pp. 222–233, 2004.
25. J. Chen, S. Sheu, and C. Yang, "A new multichannel access protocol for ieee 802.11 ad hoc wireless lans," in *IEEE Proceedings on Personal, Indoor and Mobile Radio Communications*, vol. 3, pp. 2291–2296, 2003.
26. A. Tzamaloukas and J. Garcia-Luna-Aceves, "Channel-hopping multiple access," in *IEEE ICC*, vol. 1, pp. 415–419, 2000.
27. H. So, J. Walrand, and J. Mo, "Mcmac: A multi-channel mac proposal for ad-hoc wireless networks," in *IEEE Wireless Communications and Networking Conference (WCNC)*, pp. 334–339, 2005.
28. P. Bahl, R. Chandra, and J. Dunagan, "Ssch: slotted seeded channel hopping for capacity improvement in ieee 802.11 ad-hoc wireless networks," in *ACM MobiHoc*, pp. 216–230, 2004.
29. C. Wong, R. Cheng, K. Lataief, and R. Murch, "Multiuser ofdm with adaptive subcarrier, bit, and power allocation," *IEEE Journal on Selected Areas in Communications*, vol. 17, no. 10, pp. 1747–1758, 1999.
30. P. Mähönen and M. Petrova, "Minority game for cognitive radios: Cooperating without cooperation," *Physical Communication*, vol. 1, no. 2, pp. 94–102, 2008.
31. J. Park and M. Van Der Schaar, "Medium access control protocols with memory," *IEEE/ACM Transactions on Networking (TON)*, vol. 18, no. 6, pp. 1921–1934, 2010.
32. L. Cigler and B. Faltings, "Reaching correlated equilibria through multi-agent learning," in *The 10th International Conference on Autonomous Agents and Multiagent Systems*, vol. 2, pp. 509–516, 2011.
33. I. W. Group et al., *IEEE 802.11-2007: Wireless LAN Medium Access Control (MAC) and Physical Layer (PHY) Specifications*, 2007.
34. A. Acharya, A. Misra, and S. Bansal, "Design and analysis of a cooperative medium access scheme for wireless mesh networks," in *IEEE International Conference on Broadband Networks (BroadNets)*, pp. 621–631, 2004.
35. K. Mittal and E. Belding, "Rtss/ctss: Mitigation of exposed terminals in static 802.11-based mesh networks," in *IEEE Workshop on Wireless Mesh Networks (WiMesh)*, pp. 3–12, 2006.

36. M. Vutukuru, K. Jamieson, and H. Balakrishnan, "Harnessing exposed terminals in wireless networks," in *Proceedings of the 5th USENIX Symposium on Networked Systems Design and Implementation*, pp. 59–72, 2008.
37. M. Jain, J. Choi, T. Kim, D. Bharadia, S. Seth, K. Srinivasan, P. Levis, S. Katti, and P. Sinha, "Practical, real-time, full duplex wireless," in *ACM MobiCom*, pp. 301–312, 2011.

Chapter 3
Attachment Transmission

In this chapter, we describe the overall architecture of an *Attachment Transmission* enabled communication system. The PHY architecture of attachment transmission introduces several new components. At the transmitter end, an attachment modulator is designed to enable attachment transmission when necessary. At the receiver end, we introduce an attachment detector to detect the jamming signals, an attachment demodulator to decode the attachment transmission, and an interference cancelation engine to cancel the effects of attachment transmission and recover any high-throughput content. Through the above components, we achieve the paradigm of transmitting control messages along with data packets with minimum overhead.

3.1   Overview and Design Challenges

Attachment Transmission is built on top of an OFDM-based system. It modulates control information into narrow-band signals and transmits them into air without any impact on the original data packets. Therefore, control messages and data packets can be transmitted simultaneously, and the control overhead can be minimized. The design of *Attachment Transmission* includes two components: (1) attachment modulation and demodulation and (2) attachment cancelation and data recovery. Attachment Transmission has such attractive features to avoid additional bandwidth for transmitting control messages. However, this paradigm is not easy to realize. We have encountered the following challenges:

- First, in a practical OFDM-based system, the number of available subcarriers is limited, e.g., in 802.11a/g, only 48 subcarriers are used for data transmission. Therefore, how to efficiently modulate and encode attachments remains a concern.

L. Wang et al., *Attachment Transmission in Wireless Networks*,
SpringerBriefs in Computer Science, DOI 10.1007/978-3-319-04909-0_3,
© The Author(s) 2014

Fig. 3.1 An illustration of
Attachment Transmission to
transmit control messages
with/without data packets

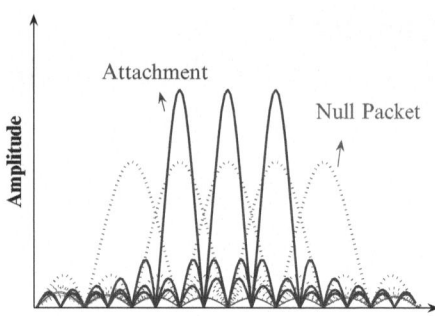

- Second, in an attachment transmission enabled communication system, the data receiver will receive a combination of attachment and data packets. It is a great challenge to separate these two components and decode data packets even when attachments are present.
- Last but not least, for the attachment listeners who want control information, it is also important for them to acquire the attachments whenever the need arises.

These challenges need to be treated carefully to achieve an attachment transmission paradigm and increase the whole system throughput. In the following section, we present the detailed design of attachment transmission to see how we deal with these challenges.

3.2   Attachment Modulation and Demodulation

In an attachment transmission enabled system, each subcarrier carries one attached signal. These attached signals across the entire channel constitute an attachment. To avoid interference with each other, each attached signal should have a bandwidth narrow enough to be included into a single subcarrier even with frequency offset. Figure 3.1 illustrates the main idea that injects attached narrow-band signals into *Null Packets* and transmits them into air. A null packet has exactly the same structure as a normal packet, except that it contains no information. The main purpose of null packets is to keep the standard frame format when transmitting an attachment.

3.2.1   Attachment Modulation

Attachments on both null packets and data packets can be modulated. We depict the procedure that modulates attachment and data packet together in Fig. 3.2. Each attachment is transmitted on one unique OFDM subcarrier with higher energy. Since the bandwidth of attachment is limited, as a payoff, its capacity is relatively small (normally only a few bits). However, this is acceptable since attachments for control information can be compressed as simple and efficient. As described later in Chap. 4,

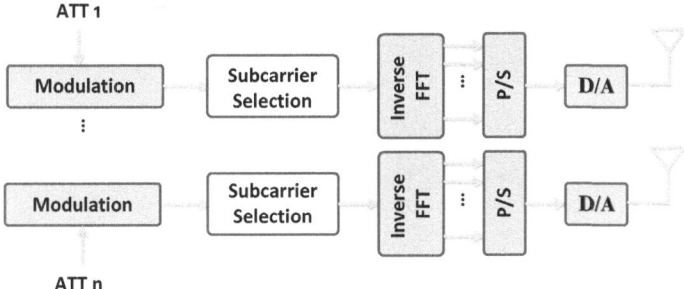

Fig. 3.2 Block diagram of attachment modulation. Each attachment is modulated on one particular subcarrier and transmitted along with them

a physical layer signaling with Binary Amplitude Modulation (BAM) is capable of modulating each attachment into only one OFDM symbol, where one attached signal on a particular subcarrier can represent certain information.

In order to allow a node to overhear attachments whenever it needs in a distributed manner, a *Cyclic Attachment* mechanism is proposed. As depicted in Fig. 3.3, each attachment is repeated on every symbol within a null packet. Then no matter which time a node starts its monitoring, the entire attachment can be retained as long as it monitors more than one symbol duration. Even when the attachment is not captured exactly from the beginning of a symbol, the missing portion can be retained from the next symbol due to the cyclic property. This cyclic property ensures that listeners can obtain control information whenever they want.

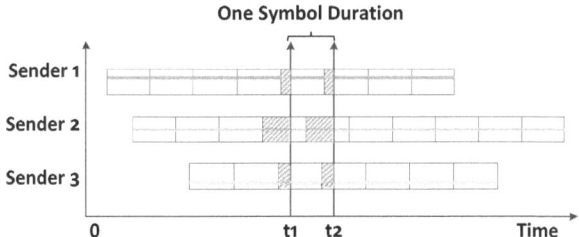

Fig. 3.3 Cyclic attachment. Each attachment is repeated on every symbol within a null packet

3.3 Attachment Demodulation

At the attachment receiver side, we adopt an energy detection-based method to detect an attached signal on a particular subcarrier. The detection principle lies in the fact that high-throughput transmissions and white noise spread their energy over

```
Attachment detector for k-th subcarrier
 1: Initialize P_avg and P_std;
 2: P_avg = RSS(s_1),  P_std = RSS(s_1);
 3: For each transmission with n symbols {s_1,s_2...s_n};
 4:   For each s_i,i ∈ [1,n]
 5:     if RSS(S_i) > P_avg + λ · P_std;
 6:       return true;
 7:     else P_avg = αP̈_avg + (1−α) · RSS(s_i);
 8:          P_std = √(αP̈_std + (1−α) · (RSS(s_i) − P_avg)²)
 9:   End if
 9:   End for
10: End for;
11: return false
```

Fig. 3.4 Pseudo-code of the attached signal detection algorithm for each of the subcarriers

the spectrum, while a narrow-band attached signal has relatively high energy level and is a kind of bursty feature. Since an attached signal has a clearly different distribution from a data signal and noise, when relatively high level energy is detected on a particular subcarrier, the attachment receiver can assume the presence of an attached signal. Although this method is simple, it is quite efficient, which can help an attachment receiver obtain an attached signal as soon as possible to interpret the corresponding control information as needed. We also notice that the detection algorithm also influences the detection accuracy and efficiency. We leave it to future research to find out more robust and efficient algorithms for attachment detection. Figure 3.4 gives the pseudo-code of the energy-based attached signal detection algorithm. We omit a detailed explanation because of the simplicity.

3.4   Attachment Cancelation and Data Recovery

In this section, we describe how to decode both attachment and data transmission. At the data receiver side, the row signals may combine both attachments and data packets. Therefore, they cannot be decoded directly. Here IC is leveraged in our design to cancel out the attached signals on each subcarrier. However, in a distributed manner with multiple concurrent data transmission in the air, it is impossible to directly adopt IC. Multiple data packets will superpose, making it challenging to record the attachment for cancelation. To address this problem, each data packet is encapsulated with a null header and null tailer. These two symbols are called "null" since ideally there is no signal except noise detected at the data receiver side. According to [1], attachments can be recorded on either header or tailer when attachments and data packets of comparable size superpose. Taking advantage of cyclic attachment, the recorded attachments contain the entire attached signal waves across all the subcarriers.

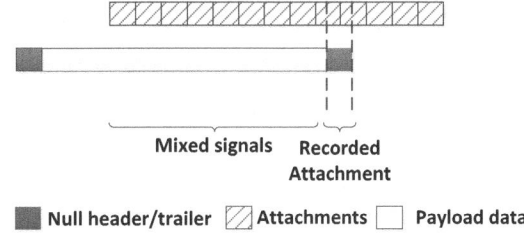

Fig. 3.5 Packet format with null header and null tailer. The recorded attachments contain the entire attached signal waves across all the subcarriers

The recorded attachment on a null header or tailer can be expressed as (Fig. 3.5):

$$y^{null}\left[t\right] = y_{attach}\left[t\right] + n\left[t\right] \tag{3.1}$$

The mixed signals in a payload data with both data and attached signals can be expressed as:

$$y^{mixed}\left[t\right] = y_{\text{data}}\left[t\right] + y_{attach}\left[t\right] + n\left[t\right] \tag{3.2}$$

where $y_{attach}\left[t\right] = H \times Attach[t]$ and $y_{data}\left[t\right] = H \times Data[t]$ are attached signals and data signals, respectively, after traversing channels to the receiver. H is the corresponding channel impulse response which can be calculated using a training sequence. $n[t]$ refers to a random complex noise. Then the original data signal can be recovered by canceling the attached signal from the mixed signal in a data symbol. So the original data symbol after the attachment cancelation can be expressed as:

$$Data_i\left[t\right] = \frac{y_i^{mixed}\left[t\right] - y_i^{null}\left[t\right]}{H} \tag{3.3}$$

After recording attachments, receivers utilize energy detection to distinguish whether a payload symbol needs interference cancelation or not. If the symbol has a bursty energy distribution, cancelation is conducted to recover that symbol and obtain the original data information.

3.5 Theoretical Analysis

To analyze the feasibility of attachment transmission in a theoretical way, we follow two principles: from a data receiver view, an attached signal cannot be too strong to corrupt the original data packet; from an attachment receiver view, the signal strength of attachment cannot be too weak to be "undetectable" in different subcarriers when multiple data packets superpose across the whole channel. Therefore, the signal strength of attachment strikes a balance between these two principles.

Table 3.1 Notations for BER calculation

Notations	Meanings
k/n	Number of information/coded bits in convolutional code
d/d_{free}	Hamming distance/free hamming distance of the convolutional code
B_d	Total number of information bit ones on all weight d paths
P_d	Probability of selecting a code word that is hamming distance d from the correct word
ρ	$\rho = W_a/W_s$ is ratio of the bandwidth of Attached signal W_a and OFDM symbol W_s

3.5.1 Reliability of Data Transmission

The Signal to Interference Ratio at the Data Receiver side (SIRD) can be expressed as E_b/N_a, where E_b and N_a are the power spectral density of the OFDM symbol and attached signal, respectively. We use Packet Reception Rate (PRR) to evaluate the quality of data transmission. As shown in Fig. 3.6, PRR has a direct connection with Bit Error Rate (BER), which is decided by the encoding/decoding scheme. Since an OFDM system applies a convolutional encoder as a channel coding scheme and a Viterbi hard decision decoder as a channel decoding scheme, we obtain an upper bound P_b on BER:

$$P_b = \frac{1}{k} \sum_{d=d_{free}}^{d_{free}+4} B_d P_d \tag{3.4}$$

P_b is calculated using Table 3.1. When d is even, P_b can be expressed as:

$$P_d = \sum_{i=\frac{d+1}{2}}^{d} \binom{d}{i} p^i (1-p)^{d-i} \tag{3.5}$$

and when d is odd, P_b can be expressed as:

$$P_d = \frac{1}{2} \binom{d}{\frac{d}{2}} p^{\frac{d}{2}} (1-p)^{\frac{d}{2}} + \sum_{i=\frac{d+1}{2}}^{d} \binom{d}{i} p^i (1-p)^{d-i} \tag{3.6}$$

p can be considered as the coded BER in an AWGN (Additive White Gaussian Noise) channel under an *Attachment* effect, with code rate $r = k/n$. OFDM adopts a Binary Phase Shift Keying (BPSK) to modulate the preamble with convolutional encoding rate $1/2$, so we first use BPSK for illustration. Each attached signal increases the noise power spectral density from N_0 to $N_0 + N_a$. Then BER for a coded OFDM subcarrier with *Attachment* is expressed as follows:

$$p = \rho \cdot Q\left(\sqrt{\frac{2rE_b}{N_0 + N_a/\rho}}\right) + (1-\rho) \cdot Q\left(\sqrt{\frac{2rE_b}{N_0}}\right) \tag{3.7}$$

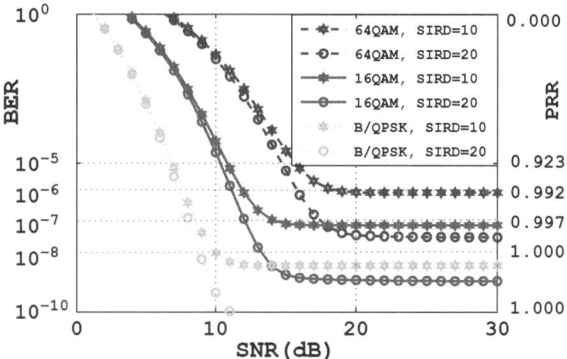

Fig. 3.6 Relationship between BER, PRR, SIRD, SNR with different modulation schemes (# of *Attachments* = 10, Packet length = 1,000 byte)

We depict Eq. (3.4) in Fig. 3.7, which shows the relationship between PRR, BER, SIRD, and SNR using different modulation schemes. It is noted that the typical working range of WLAN is from 20 to 30 dB for wireless networks [2]. With a reasonable number of *Attachments* as 10, BER is smaller than 10^{-7} with B/QPSK and 16 QAM, resulting in a PRR of 99.9 %. Even with 64 QAM, we can achieve a PRR of 99.2 %, which is sufficient for current 802.11 specifications. Therefore, Attachment Coding is quite harmless in relation to the original data transmission.

3.5.2 Feasibility of Attachment Transmission

We define the Signal to Interference Ratio at *Attachment* Receiver side (SIRA) as N_a/E_b. Therefore, the received signal sample of an intended sender can be represented by the following expression:

$$y(m) = \sum_{i=1}^{n} h_i(m)[A_i(m) + D_i(m)] + w(m) \tag{3.8}$$

where m denotes the sample index and $h_i(m)$ denotes the impulse response of the ith channel. Without loss of generality, we assume the transmission channel is an AWGN channel, that is, $h_i(m) = h_0 = 1$. $A_i(m)$ and $D_i(m)$ are the attached signal and data signal of the ith channel, with zero-mean and variance of N_a and E_b, respectively. $w(m)$ denotes a complex Gaussian Noise with zero-mean and variance of N_0. According to [3], the probability of missing an *Attachment* when one is present on a certain subcarrier P_{miss} is:

$$P_{miss}(\lambda) = Pr(\frac{1}{M} \sum_{m=1}^{M} |y(m)|^2 < \lambda) \tag{3.9}$$

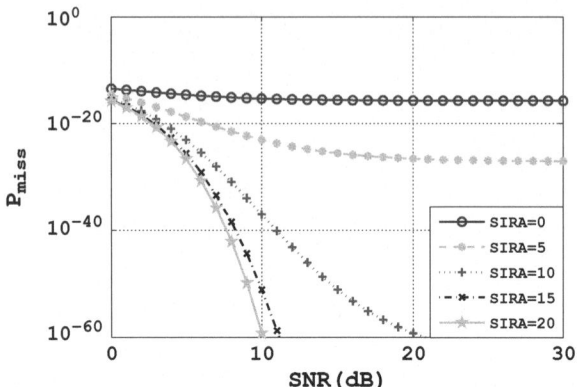

Fig. 3.7 Probability of missing an attached signal under different SNR and SIRA conditions

where M is the number of samples and N is the maximum number of neighbors among a node. The threshold level for energy detection, λ, should be at least larger than $N \cdot E_b$, so that the attached signal can be detected using an energy detection mechanism. We depict Eq. (3.9) in Fig. 3.7 to see the probability of miss detection P_{miss} under different SIRA and SNR. Generally, P_{miss} is acceptable in a typical wireless working range, with values below 10^{-25}. Therefore, we can conclude that the signal strength of an *Attachment* cannot be much larger than the signal strength of a data symbol. In this way we can ensure the performance of both data transmission and attachment detection.

3.6   Performance Evaluation

In this part, we conduct real-time experiments to evaluate the feasibility of attachment transmission. The evaluation comprises two aspects: (1) whether data transmission can be reliability decoded in the presence of attachment transmission and (2) whether we can successfully detect attachment transmission and obtain the information from it. These two aspects are consistent with theoretical analysis in Sect. 3.5. We first outline the system implementation.

3.6.1   System Implementation

We utilize GNU radio testbed for our experiments and implement Attachment Coding using Software Defined Radios (SDRs). The Universal Software Radio Peripheral 2 (USRP2) is used as RP frontend. Our testbed consists of 8 USRP2

Fig. 3.8 Experimental environment of our GNU testbed (three sets of the four nodes' locations are illustrated as an example)

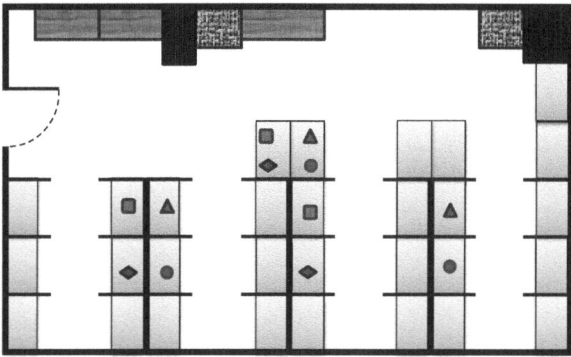

Table 3.2 Experimental configuration parameters

Parameters	Values	Parameters	Values
SIFS	16 μs	DIFS	32 μs
Symbol time	16 μs	Slot time	9 μs
CW_{min}	16	CW_{max}	1,024
Packet length	1,460 bytes	Basic data rate	6 Mbps

nodes with RFX2400 daughterboards operating in the 802.11 frequency range. Our implementation uses BPSK as the modulation scheme. Unless otherwise specified below, we use the default configuration as shown in Table 3.2. Specifically, we use a bandwidth of around 2 MHz and split it into 52 subcarriers. These changes are made since we want to make the inter subcarrier spacing comparable to 802.11 (0.3125 MHz) while still maintaining the normal transmission of USRP2, which is limited by the hardware itself. We also make SIFS and DIFS longer to ensure that during attachment sense, a sender can overhear a whole OFDM symbol [4]. All of our experiments run on the 2.425 GHz. We then still follow the two evaluation principles discussed in previous subsection.

3.6.2 Reliability of Data Transmission

To evaluate the reliability of data transmission under the impact of Attachment Transmission, we first measure the decodability of the data receiver with and without attachment transmission. Here a four-node setting is configured, that is, two nodes for data transmission and two nodes for attachment transmission. As shown in Fig. 3.8, the data receiver is in the transmission range of both the data sender and the attachment sender. We let the data sender transmit normal packet to the data receiver, and simultaneously the attachment sender transmit *Attachments* to the attachment receiver. We compute the PRR at the data receiver side under various SNRs, first without jamming, then with jamming. Each run transfers 2,500 packets, and for each value of SNR, the experiment is repeated 10 times.

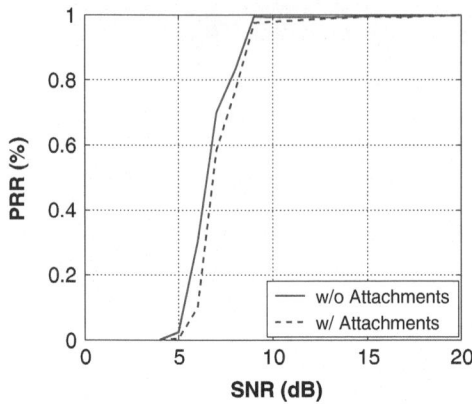

Fig. 3.9 Decodability of data transmission with/without attachment transmission under different SNRs. The results reveal that attachment transmission has little influence on the data transmission

We plot the PRR of data receiver with/without attachment transmission as a function of the received SNR at the data sender side from [4, 20] dB in Fig. 3.9. We can see that when the SNR exceeds certain threshold, that is, 10 dB, the PRRs with Attachment Transmission are almost the same as those without attachment transmission. There is a little performance degradation when the SNR is smaller than 10 dB, which, however, can be acceptable, since the typical working range of SNR region for 802.11 is 10–30 dB [5], verifying that attachment transmission does not have any influence on the original data transmission. It is noticed that these experimental results are not as good as in the theoretical analysis in Sect. 3.5. This is due to two reasons. On the one hand, USRP has certain limitations in strict timing and accurate sampling due to software-defined signal processing. On the other hand, our implementation runs in a public user-space in the unlicensed 2.4 GHz range. Therefore, there must be some external interferences that cannot be avoided.

In the next step, we evaluate the impact of the number of concurrent attachment transmissions on the decodability of the data receiver. We use a similar setting to evaluate the PRR of the data receiver but with different numbers of concurrent attachment senders varying from 1 to 6. We assign each attachment sender a unique subcarrier for attachment transmission in this experiment, which are Subcarrier 1, 3, 5, 7, 9, and 17.

We plot the PRRs under different numbers of attachment transmissions with SNR 11 and 15 dB in Fig. 3.10. We can see that all the performance losses are under 10^{-2}, even with six concurrent attachment transmissions, thus the performance loss is relatively small and acceptable. It is noticed that according to the theoretical analysis in Sect. 3.5, the performance loss is expected to increase as the number of concurrent attachment transmissions increases. However, the experimental results show that the performance loss varies randomly under different number of concurrent attachment transmissions. This difference between the practical and the theoretical results may be due to the processing capability of USRP2 hardware.

Fig. 3.10 Impact of number
of concurrent attachment
transmission under different
SNRs

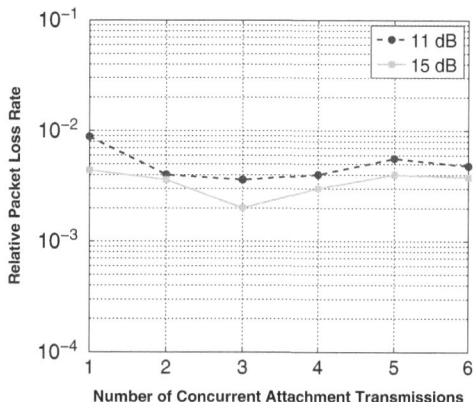

3.6.3   Feasibility of Attachment Transmission

To evaluate the performance of attachment transmission, we measure the detection accuracy at the attachment receiver side, that is, whether the attachment receiver can correctly detect an *Attachment* and decode the attached control information. There are two aspects that influence the detection accuracy: Miss Detection Rate (P_{miss}) and False Alarm Rate (P_{false}), both of which result in decoding failure. Here we still use a four-node setting, that is, two nodes for data transmission and two nodes for Attachment Transmission. We let attachment sender keep transmitting *Attachments* in the presence of data transmission, then PRR is computed under various SNRs of the attached signal ranging from [8, 20] dB. Each run transfers 2,500 packets, and for each value of SNR, the experiment is repeated 10 times.

According to the theoretical analysis in Sect. 3.5, we expect that the P_{false} is very small. From the experimental results we find that there is almost no P_{false} for all runs. Therefore, we only plot the results of P_{miss} in Fig. 3.11. We can see that when SNR >13 dB, P_{miss} can be controlled within 1 %, which results in a detection accuracy of

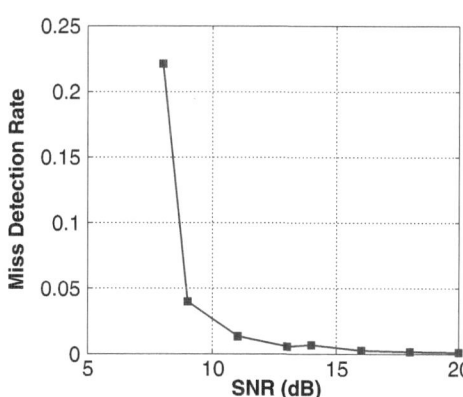

Fig. 3.11 Miss detection rate
of attachments under different
SNRs

more than 99 %. It is noted that the detection algorithm will also have an impact on the results. Therefore, we consider designing a more precise detection algorithm as one of our future works.

References

1. M. Vutukuru, K. Jamieson, and H. Balakrishnan, "Harnessing exposed terminals in wireless networks," in *Proceedings of the 5th USENIX Symposium on Networked Systems Design and Implementation*, pp. 59–72, 2008.
2. M. Souryal, L. Klein-Berndt, L. Miller, and N. Moayeri, "Link assessment in an indoor 802.11 network," in *IEEE Wireless Communications and Networking Conference (WCNC)*, vol. 3, pp. 1402–1407, 2006.
3. Y. Lee and D. Oh, "Energy detection based spectrum sensing for sensing error minimization in cognitive radio networks," *International Journal of Communication Networks and Information Security (IJCNIS)*, vol. 1, no. 1, 2011.
4. K. Tan, J. Fang, Y. Zhang, S. Chen, L. Shi, J. Zhang, and Y. Zhang, "Fine-grained channel access in wireless lan," in *ACM SIGCOMM Computer Communication Review*, vol. 40, pp. 147–158, 2010.
5. M. Souryal, L. Klein-Berndt, L. Miller, and N. Moayeri, "Link assessment in an indoor 802.11 network," in *IEEE Wireless Communications and Networking Conference (WCNC)*, vol. 3, pp. 1402–1407, 2006.

Chapter 4
Applications to Classic Problems

In this chapter, we present several typical applications that leverage attachment transmission to solve the above-mentioned classic problems. With the assistance of attachment transmission, a variety of MAC layer protocols are designed, which can guide mobile stations to make better access decision through the PHY layer information they need. These applications include: *Harmless Attachment* that uses attachment transmission to benefit multiple access in WLANs in the infrastructure mode [1]. *Attachment Learning* that helps mobile stations to learn allocation strategy by themselves. After the learning stage, mobile stations can achieve a TDMA-like performance, where stations can know when exactly to transmit on which channel without further collisions. *Attachment Sense* that identifies hidden and exposed terminals in ad hoc networks. The self-attached control provides accurate channel usage information (CUI) in real time, thus guides mobile stations to make the right channel access decision fast and accurate. In the following sections, we will present how we utilize attachment transmission to solve the above-mentioned problems.

4.1 Harmless Attachment for Multiple Access in WLANs

In this section, we show how to use attachment transmission to benefit transmissions in WLANs in the infrastructure mode. Traditional wisdom leverages RTS/CTS to coordinate among clients and AP. However, as illustrated in Fig. 4.1, RTS/CTS consumes rather massive overhead, and thus greatly degrades the performance. Here, we leverage attachment transmission to solve this problem. The new communication architecture proposed here is called Harmless Attachment with the core idea in Fig. 4.2, where the high-throughput transmission is used for application data traffic and the attachment is used for control message delivery.

L. Wang et al., *Attachment Transmission in Wireless Networks*,
SpringerBriefs in Computer Science, DOI 10.1007/978-3-319-04909-0_4,
© The Author(s) 2014

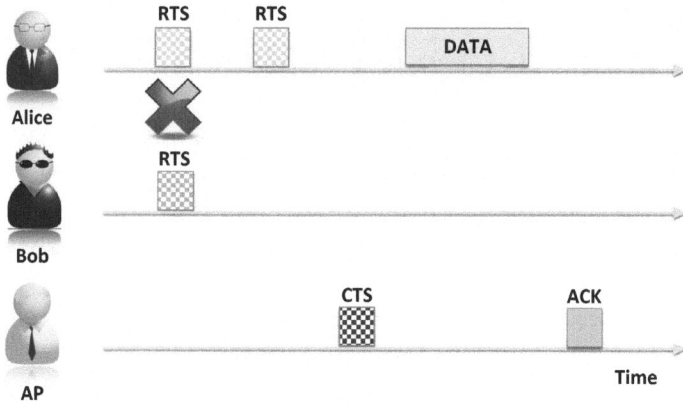

Fig. 4.1 An illustrative example of traditional multiple access mechanism. The contention resolution overhead greatly degrades the performance

4.1.1 Harmless Attachment Overview

To reduce the multiple access overhead in infrastructure-based WLANs, we propose Harmless Attachment, where the access request is modulated into attachments. Consider a simple transmission scenario with three clients Alice, Bob and Dave, and an AP, as illustrated in Fig. 4.2. Suppose Alice obtains a high-throughput channel for the next transmission. Alice is in normal mode which transmits the content in the traditional way (High-throughput Tx. is exactly the same as OFDM Tx.). The others (Bob and Dave) will then turn to the attachment transmission mode and attempt to use the attachment transmission. Each client in attachment transmission mode will select a unique subcarrier assigned by AP and send attached signals when Alice is sending. These attached signals carry the attached information from the attachment transmission clients, combine Alice's signal in the air and are received by the AP. At the receiver end, the AP first applies the attachment detector to determine whether any attached signals from attachment transmission clients exist. These attached signals are then analyzed and decoded to recover attachment transmissions (from Bob or Dave). In the meantime, the Interference Cancelation technique [2] is applied to cancel the attached signals and recover the original data for the high-throughput client (Alice).

4.1.2 System Architecture

Consider the single AP scenario. Upon receiving a data packet from a client, AP first decodes both the high-throughput and attachment transmissions. Then the high-throughput content is delivered to the upper layer application directly while the

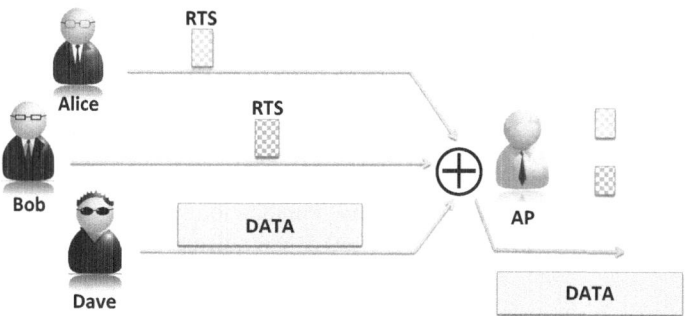

Fig. 4.2 An illustrative example of an attachment transmission enabled communication system. The control messages are transmitted together with data packets, without degrading the effective throughput of the data transmission

attachments are collected for coordination purpose. These attachments carry the transmission requests from the clients and can be further used to build a potential sender list. By having this list, the AP is responsible for whole channel coordination and assigns the next sender. Specifically, the AP attaches the senders' IDs in order in the ACK and broadcasts it. At the client end, by receiving the ACK from AP, the client can check its order in the sender list and determines whether it is the next sender of the high-throughput channel. Then the next data transmission continues. It will be similar when the transmission is from the AP to the client due to that each client knows its sending order.

In addition, clients may join and leave. At the initialization step, the AP is responsible for allocating the subcarriers to the existing clients in the network. Afterward, a client being inactive for too long time is automatically kicked out by the AP. To the contrary, a newcomer should first listen to the AP's broadcast ACK packet (indeed, the ACK is for other clients). This packet carries the sub-channel utilization information and the newcomer simply selects a random unused subcarrier to deliver its request.

4.1.3   Points of Discussion

Some practical issues in the harmless attachment are discussed in this section. For the issues raised below, we broadly describe potential approaches to handle them. However, it leaves an exhaustive discussion in future works.

4.1.3.1   Multiple AP Scenarios

Besides the infrastructure mode, here we consider a network with multiple APs in different collision domain. Actually as we mentioned before, we can solve this problem by assigning them orthogonal channels. However, if the channels are not enough and should be non-orthogonal, how to coordinate the clients in the overlapped region becomes a challenge. In that case, we reserve one subcarrier for the contention of these clients. The clients in the overlapped region between two APs will contend the channel through the reserved subcarrier. And the contention procedure is similar to the traditional CSMA. Then the associate AP will also use this subcarrier to send back the ACK. Since we reserve one subcarrier for the clients in the overlapped region, the performance degradation is $1/48 \approx 2\%$. However, when comparing with the traditional CSMA, we still have significant throughput gain.

4.1.3.2 Priority of Transmission

For some latency sensitive transmission, the priority of the transmissions will significantly affect the network performance. For example, SSH only needs to send several bits information but has to wait for a long time based on the traditional contention methods. Especially for some real-time applications, the delay is intolerable. However, in the traditional approaches, all nodes need to contend the channel before transmission and the workload information of the data traffic is not taken into account for priority scheduling. In harmless attachment, the load indicator [3] of the packets can be delivered as the attached information. Then, we can design some smart scheduling algorithms based on such information to reduce the network latency.

4.1.4 Performance Evaluation

We implement a simulator to understand the performance of harmless attachment primarily under a single AP network with varying number of total clients. In order to focus on the performance on the channel utilization by each method, we assume that the packet reception failure is only caused by the collisions and the network is saturated. In this subsection, we mainly compare the performance of harmless attachment with CSMA/CA [4], which is used by the current IEEE 802.11 Standard.

Our simulations model the CSMA MAC. For scheduling of harmless attachment, we just simply use Round Robin as an example to demonstrate its performance. For the evaluation of harmless attachment, Eq. (4.1) gives a simple model for harmless attachment's throughput efficiency.

$$E_{harmless\ attachment} = \frac{t_{data}}{t_{preamble} + t_{data} + t_{ACK}} \tag{4.1}$$

Unless otherwise stated, the default packet size is 1,500 bytes, which is around the maximal transmit unit (MTU). Table 4.1 summarizes the configuration

Table 4.1 Experimental configuration parameters

Parameters	Values	Parameters	Values
SIFS	$10\,\mu s$	DIFS	$28\,\mu s$
PIFS	$19\,\mu s$	Slot time	$9\,\mu s$
Preamble	$20\,\mu s$	Symbol time	$4\,\mu s$
CW_{min}	16	CW_{max}	1,024

parameters used in our simulators. Specially, for the simulation of high data rates with 802.11n, the total number of subcarriers is set to 114, of which 108 are used for data transmission and 6 for equalization.

Figure 4.3 plots the throughput gain of harmless attachment over 802.11 as a function of total clients under different data rates. It shows that harmless attachment outperforms 802.11 CSMA MAC at all cases and its throughput gain is significant, e.g., when the data rate is 54 Mbps, the relative throughput gain over 802.11 a/g is up to around 72 %. Figure 4.4 further illustrates the throughput gain over 802.11n in higher data rates. Compared to 802.11n at a data rate 600 Mbps, harmless attachment can even achieve a gain up to 200 %. This significant improvement is due to two reasons. First, harmless attachment eliminates the coordination overhead in each transmission while the proportion of coordination overhead in CSMA increases as the data rate increases. The second reason is due to that harmless attachment is collision free while the collision probability of CSMA in IEEE 802.11 increases with the number of clients. Thus harmless attachment has better utilization of the channel.

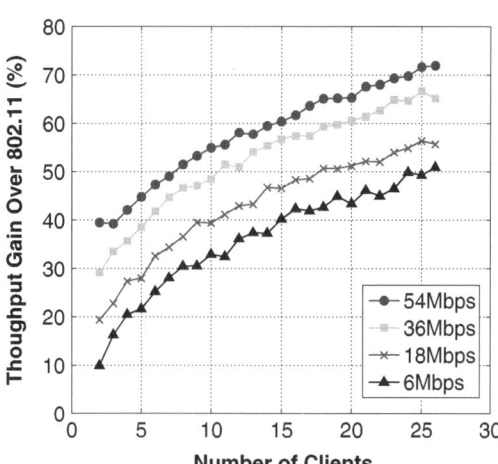

Fig. 4.3 Performance gain of harmless attachment over 802.11 a/g under different number of clients

Fig. 4.4 Performance gain of
harmless attachment over
802.11n under different
number of clients

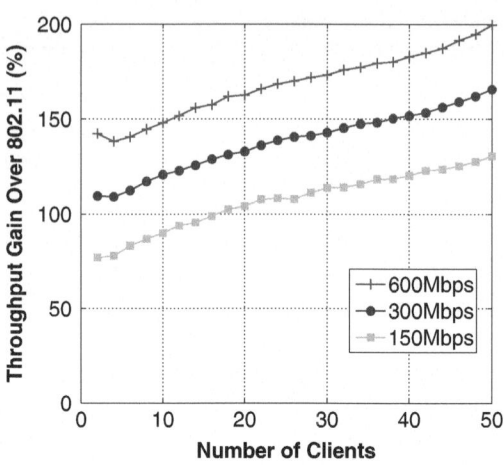

4.1.4.1 Summary of the Results

Our experiments and simulations reveal the following:

- We have prototyped the implementation of harmless attachment on a testbed consisting of 8 USRP2 nodes to prove the feasibility of harmless attachment. Harmless attachment provides a high decodability around 100 % and achieves a 99 % detection accuracy when SNR is greater than 13 dB.
- Harmless attachment provides significant throughput gain over tradition 802.11 CSMA MAC by up to 200 %.

Moreover, the feasibility results in this paper are derived from laboratory experiments, without node mobility. Harmless attachment under harsh conditions remains for our future work.

4.2 Attachment Learning for Multichannel Allocation

Multichannel environments such as OFDM-based systems call for efficient channel allocation protocols. For a centralized network, such protocols have already been well studied, since there always exist certain authorities (e.g., Access Points) who can directly designate the channel allocation for all the stations in that network [5]. However, in distributed networks, since there are no such authorities, channel allocation simply relies on coordination among stations (cooperative) or history knowledge of themselves (noncooperative). The former retails a rather high overhead, and the latter method has relatively low accuracy. Thus neither of them can achieve the desired utilization.

4.2.1   Attachment Learning Overview

Many existing works try to solve multichannel allocation problem in a game theoretical way [6–8]. The aim is to achieve NE, which guarantees the best payoff for all nodes. However, not all the games have desirable equilibrium structure, and the fairness among players cannot be ensured [9]. Based on the above observation, we conclude that, for a multichannel allocation game in distributed networks, we need a noncooperative scheme with efficient NE and ensured fairness. Specifically, without coordination, stations are better to learn an efficient access strategy by themselves. Therefore, we adopt a Correlated Equilibrium (CE) instead of NE. Figure 4.5 demonstrates the basic idea of CE, which is a probability distribution over the joint strategy profiles in the game [10]. It assumes a correlation device for all the players, which samples the probability distribution and recommends an action for each player. When none of the players can increase its payoff by deviating from the recommended action, the distribution reaches CE. Since the correlation device services as an authority, it can both ensure NE and fairness.

Fig. 4.5 An illustrative example of Correlated Equilibrium. A correlation device samples the probability distribution and recommends an action for each client

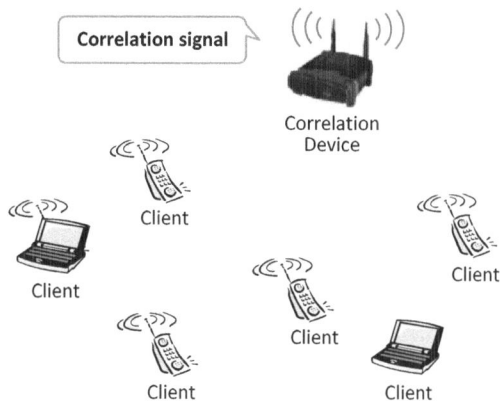

However, it is nontrivial to achieve CE in distributed networks. We encounter the following challenges: First, the coordination device is not available in distributed networks. We need to find an alternative that serves the same purpose. Second, a complete MAC protocol is required to fully utilize a correlation signal to achieve CE for multichannel allocation.

To address the above challenges, we propose Attachment Learning (AT-Learning), which is a cross-layer design that consists of identifier attachment in PHY layer and Identifier Learning in MAC layer. Identifier attachment utilizes attachment transmission 3 to provide a coordination signal for MAC layer. Then Identifier Learning guides stations to learning a channel allocation strategy themselves. These two components together contribute to achieving CE among stations.

By exploring attachment transmission, senders are able to inject specially designed jamming signals as identifiers on their own data packets and transmit these two types of signals simultaneously in the same channel. These identifier signals serve as the above-mentioned coordination signal. This idea is illustrated in Fig. 4.6. By implementing a secondary radio for dedicated listening, each station can overhear all the identifiers across the entire channels using Jamming Detection. Meanwhile, the receivers manage to remove the attached jamming signals from the received data stream by Interference Cancelation, and thus can successfully recover the original data packets. This attachment transmission helps us self-generate a coordination device without further consuming any channel resource.

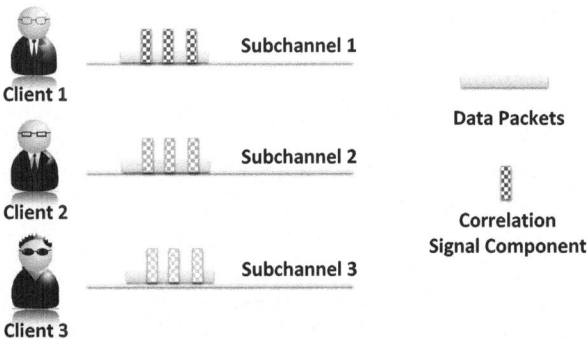

Fig. 4.6 An overview of attachment learning for multichannel allocation problem. Clients inject specially designed jamming signals as identifiers on their own data packets and transmit these two types of signals simultaneously in the same channel

After gathering the coordination signal from the PHY layer, we propose an Identifier Learning in the MAC layer for distributed channel allocation without coordination. Identifier Learning helps stations within the same collision domain learn an efficient allocation strategy based on each value of the observed coordination signal. Specifically, time is slotted into transmission rounds. Each station maintains a strategy table, mapping each coordination signal to an available channel. Before each transmission round, stations consult their strategy tables and make channel access decisions according to the coordination signal value observed from the last transmission slot. If a transmission failed, the mapping in the strategy table will be adapted, otherwise it remains unchanged. Through theoretical analysis and simulation we argue that this MAC protocol can achieve CE of a multichannel allocation game. It can also guarantee fairness among stations.

4.2.2 Resource Allocation Game

Slot allocation in OFDMA-based systems for multi-Client access can be formulated into a resource allocation game. A particular frequency band in a certain period of time is considered as a slot. Each slot is the basic resource unit for channel allocation. Stations simply contend for a slot to transmit data. In this section, we first give a brief introduction to a resource allocation game as our problem formulation. Then we will see how a learning-based algorithm is proposed to handle this resource allocation game.

4.2.2.1   Problem Formulation

Resource allocation game is defined as a game between M clients and S channels. These clients are the players who always want to obtain transmission access to one of the channels, as to maximize their payoffs. Here we assume $M \geq S$, since in practice, we always encounter the case that there are more clients than channels. Also, channel is slotted. Each access from a client gains one exclusive slot for its transmission. In distributed systems, clients are independent from each other. It is extremely difficult for them to achieve Correlated Equilibrium of allocated resources without coordination. Learning-based algorithm [9] is an optimal solution for clients to learn a "steady state," with efficient NE and ensured fairness. We propose that a randomly chosen integer exists, which is independent from the channel condition and can be observed by every client from time to time. This random integer serves as a "stupid" coordination signal. The "smart" clients learn which action they should use for each value of the coordination signal. Specifically, each client maintains an access strategy table (AST) and each coordination signal is mapped to a single channel. Clients observe the common coordination signal before each round and then decide which channel they will use in this round. According to the outcomes of their transmissions (success or failure), they will decide whether to change their strategies or not. In the next subsection, we will show the detailed operations of this learning-based algorithm.

4.2.2.2   Algorithm Description

We define the set $\mathbf{M} = \{1, \ldots, M\}$ as M number of clients, and set $\mathbf{S} = \{1^i, \ldots, S^i\}$ as S number of channels. For each subset $\mathbf{s}^i$ included in set $\mathbf{S}$, time is divided into i number of even slots. The coordination signal has signal space of $\mathbf{C} = \{1, \ldots, C\}$ and each value remains stable at the beginning of each time slot. Each client stores an AST. AST of client m is defined as $f_m : \mathbf{C} \rightarrow \mathbf{S} \cup \{0\}$. This table simply maps each coordination signal into an exclusive channel or zero, which exactly indicates the access action in every time slot. Specifically, in time slot t, if the observed coordination signal $f_m(c_t) = 0$, client m does not have channel access authority and should defer its transmission in time slot t. Otherwise, if $f_m(c_t) > 0$, client m can have access to channel $f_m(c_t)$ and conduct transmission immediately.

AST is initialized as follows: for each coordination signal $c_0 \in \mathbf{C}$, client m uniformly chooses one channel from $\mathbf{S}$ and assigns it a coordination signal c_0. This randomized manner can ensure fairness among clients, since they have equal chance to access each channel.

When transmission starts, clients access channel according to their ASTs. Since initially AST is randomized, collisions are unavoidable. Also, there might be some channels remain vacant. Therefore, clients adapt their strategies in the following two phases: transmission and monitoring.

Transmission

At time slot t, if $f_m(c_t) > 0$, client m tries to transmit over channel $f_m(c_t)$. After transmission, it observes the outcome of its transmission:

- If the transmission succeeded, client m keeps mapping channel $f_m(c_t)$ to this coordination signal r_t and AST remains unchanged.
- If the transmission failed, client m assumes that collision might happen. So it sets $f_m(c_t) = 0$ with probability P_{defer}, which means that it should defer transmission for coordination signal c_t to avoid further collision.

Monitoring

At time slot t, if $f_m(c_t) = 0$, client m defers its transmission in this time slot. Meanwhile, it chooses a channel $s_i'(t) \in \mathbf{S}$ to monitor the activity.

- If $s_i'(t)$ is free, client m sets $f_m(c_t) = s_i'(t)$ for coordination signal c_t.
- If there is any transmission on $s_i'(t)$, client m keeps AST unchanged.

The above learning-based algorithm adopts a constant defer mechanism, where clients defer with the same probability P_{defer} when collision happens. However, when encountering a collision, it is not necessary for all the clients to defer with the same probability. So we amend the constant mechanism to get a couple of variants. First, we let client m defer with the probability $P_{defer} = |f_m|/c$, where $|f_m|$ refers to the degree of AST (the number of available channels contained in that AST). Then collided parties will be more likely to have different access decisions after the AST adaptation. This is called linear defer mechanism, where clients defer according to $|f_m|$. Furthermore, we let client m that has the lowest $|f_m|$ keep AST unchanged when collision happens, and other collided parties defer transmission according to $P_{defer} = |f_m|/c$. This greedy protocol guarantees that at least one client will transmit in the following round. However, it requires clients to obtain others' $|f_m|$s through coordination, and thus it is more complex.

4.2.2.3 Algorithm Analysis

To evaluate the feasibility and reliability of the proposed learning-based algorithm, we use two metrics: convergence time to CE and fairness among clients. The convergence time is the estimated number of steps that all the clients can achieve to the "steady state," that is, there are no collisions, and in every channel for every signal value, some client transmits. While fairness among clients is to see whether clients have equal chance to access all the channels after stability.

Convergence Time

To see how fast this learning-based algorithm can converge to a CE, we divide the calculations into several steps. First, we prove the convergence for simple cases, where $S = 1$ and $C = 1$, along with $S = 1$ and $C > 1$. Then we show the general case that $S \geq 1$ and $C \geq 1$. We all assume there are M clients. According to the calculation in [9], we can obtain the following theorems.

Theorem 4.1. *For $S = 1$, $C = 1$, and $0 < p < 1$, the expected number of steps to converge to a pure-strategy Nash equilibrium of the resource allocation game is*

$$O\left(\frac{1}{p\,(1-p)}\log M\right)$$

.

Theorem 4.2. *For $S \geq 1$, $C = 1$, and $0 < p < 1$, the expected number of steps to converge to a pure-strategy Nash equilibrium of the resource allocation game is*

$$O\left(S\frac{1}{1-p}\left[\frac{1}{p}\log M + S\right]\right)$$

Theorem 4.3. *For $S \geq 1$, $C \geq 1$, and $0 < p < 1$, the expected number of steps to converge to a pure-strategy Nash equilibrium of the resource allocation game for every $c \in C$*

$$O\left(C^2 S\frac{1}{1-p}\left[\frac{1}{p}\log M + S\right]\right)$$

Therefore, the learning-based algorithm converges in expected polynomial time in the number of clients and channels to reach a steady state (an efficient pure-strategy NE) for all the cases.

Fairness

Another metric for evaluation is fairness among clients after converging to an efficient CE. We define the number of slots won by a client i across all the time slots as a random variable X_i. This variable follows a binomial distribution, denoted by $X_i \sim B(n, p)$, where $n = C$. Since clients have independent decisions in each time slot, every client has an equal chance to win. For M clients and S available channels, the probability that a client can win a given slot is $p = \frac{S}{M}$. For a random variable X_i, we use *Jain index* [11] to measure fairness:

$$J(X) = \frac{(E[X])^2}{E[X^2]} \tag{4.2}$$

Since $E[X] = C \cdot S/M$, we obtain the following equations:

$$E[X^2] = \left(C \cdot \frac{S}{M}\right)^2 + C \cdot \frac{S}{M} \cdot \frac{M-S}{M} \tag{4.3}$$

Therefore, the *Jain index* $J(X)$ can be interpolated as:

$$J[X] = \frac{S \cdot C}{S \cdot C + (M-S)} \tag{4.4}$$

An allocation is considered fair if $J(X)$ is close to 1, which means that all the clients have close possibilities to win equal number of slots across all the time slots. For any value of S, it holds that $\lim_{M \to \infty} \frac{M}{S \cdot C} = 0$ if $C = \omega(M/s)$ (C is much larger than M/s, and we assume $M \geq S$ as mentioned above). Then we can obtain $\lim_{M \to \infty} J(X) = 1$, which indicates that if we choose a relatively large C, the resource allocation becomes fairly equitable as M goes to infinity. Therefore, the fairness increases as the signal space of C increases.

4.2.3 System Architecture

In this section, we will provide the design of an Attachment Learning system for multichannel allocation in OFDMA-based systems. First, an overview of AT-Learning is given along with the design challenges. Detailed modules of AT-Learning are then presented to see how we address these challenges, including PHY layer Identifier Attachment and MAC layer Identifier Learning.

4.2.3.1 Protocol Overview

A Learning-based algorithm that achieves CE gives us an insight into the design of slot allocation in OFDMA-based systems. However, it holds several challenges for implementation.

- First, it is nontrivial to provide a common coordination signal for all the stations within the same collision domain. The simple noise suggested in [10] is not feasible due to the uncertainty of the wireless channel. We need to provide a reliable and feasible coordination signal.

Fig. 4.7 Coordination signal vector **C**. There are totally five subchannels. At time t_1, the attached signal on each subchannel is 1, 0, 1, 4, and 3. So the coordination signal vector $\mathbf{C}_{t_1}$ is $\{1, 0, 1, 4, 3\}$

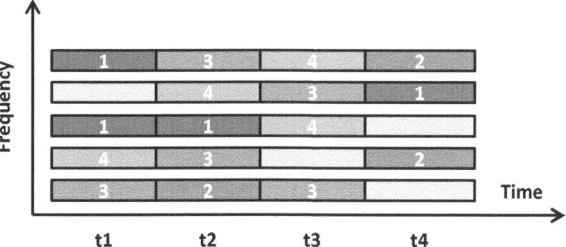

- Second, since the observation stage for the coordination signal before each transmission time slot is rather expensive, we need to minimize the observation overhead.
- Third, in multichannel scenario, the sender and receiver have to negotiate before the transmission, which makes it much more complex to implement the learning-based algorithm.

To address these challenges, we proposed an identifier attachment-based coordination signal in distributed OFDMA based networks. This coordination signal $\mathbf{C}_t$ is an S dimensional vector, where S is the number of subchannels. Each component of $\mathbf{C}_t$ is a narrow-band signal attached on a data symbol in each subchannel. As illustrated in Fig. 4.7, there are totally five subchannels. At time t_1, the attached signal on each subchannel is 1, 0, 1, 4, and 3. So the coordination signal vector $\mathbf{C}_{t_1}$ is $\{1, 0, 1, 4, 3\}$. A second radio is adopted for each station to listen on all the subchannels for the coordination signal. Therefore, all the stations within the same collision domain can observe the same coordination signal vector from time to time. Also, since subchannels are slotted, stations contend for one subchannel in each time slot. In this paper we only consider single-cell OFDMA-based systems. The case for multiple collision domains raises other problems and thus is remained as a future research. As previously shown, when an optimization problem is deployed in an OFDMA-based system, time considerations are crucial. Therefore, we assume that the problems are handled in a framework of frames.

4.2.3.2 Attachment Learning

The MAC layer protocol is built on top of the PHY layer information. Complying with the major standard of OFDMA-based networks, there are some necessary assumptions listed below:

1. There are S adjacent subchannels of interest. Each of them has the same number of subcarriers r.
2. Time is divided into even time slots. Each time slot t is used for one transmission round, including data packet and ACK feedback.

Table 4.2 An illustrated example of strategy table of station i

C_t	1	2	$\cdots$	$C-1$	C
$f_i(C_t)$	S_2	0	$\cdots$	S_1	S_3

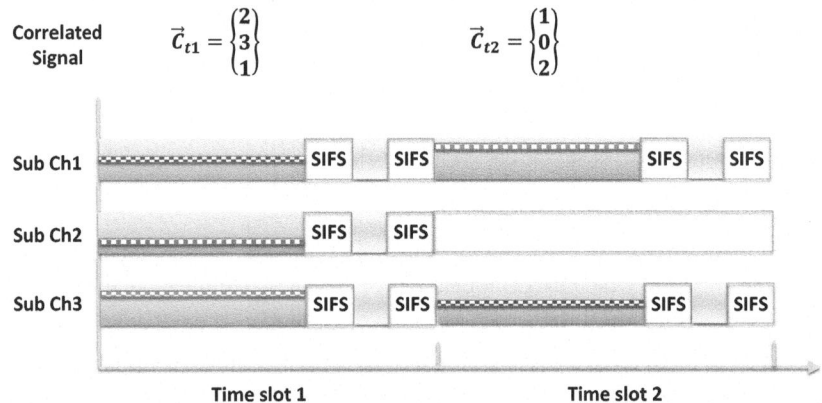

Fig. 4.8 Multiple transmissions along with the attached signals in OFDMA-based systems

3. Stations get implicitly synchronized when the channel becomes idle, as stated in [12]. They all transmit during one time slot.
4. Each station equips with two half-duplex antennas, one is for data transmission and the other is for coordination signal sensing.

Each station maintains an online strategy table, which stores the mappings from coordination signal space to available channel space. Since we have S subchannels and each of them with r subcarriers, the coordination signal space is r^S. We can further modify this space by balancing the number of subchannels and subcarriers. Initially a strategy table is constructed by randomly distributing each available subchannel across the whole coordination signal space. Table 4.2 depicts an example for a particular station to initialize its strategy table. We can see that subchannels are mapped stochastically to each coordination signal. This randomness ensures that stations have equal chance to access channels across all available coordination signals.

Stations access subchannels according to the $\mathbf{C_t}$ observed in the previous time slot. Specifically, at the beginning of slot $(t+1)$, they check $f_i(\mathbf{C_t})$ for a decision, where $\mathbf{C_t}$ is gathered in time slot t. To start transmission at time slot 1, since there is no transmission before, no coordination signal is at hand. Stations simply treat $\mathbf{C_0}$ as 0. As illustrated in Fig. 4.8, when transmissions are being conducted, components of $\mathbf{C_t}$ are attached on data packets. All the stations sense $\mathbf{C_t}$ and store it for the purpose of access decision in the next time slot. Stations that do not transmit just keep a receiving mode. Since an antenna can receive all the signals across the whole bandwidth, the sender/receiver negotiation is avoidable. After each transmission, the sender adapts its strategy table if there is no ACK from its receiver (in this case we assume collision happens), by setting its mapping for C_t from $f_i(C_t)$ to 0. Otherwise if ACK is received, the sender remains its mapping for C_t unchanged. Meanwhile, other stations who do not conduct transmission randomly choose one subchannel S_j to monitor. If the subchannel is free, they set their mappings for C_t as S_j.

4.2.4 Performance Evaluation

Since USPR2 has a latency constraint, we cannot use it to conduct real-time evaluation of the whole AT-Learning system. Therefore, we implement a self-defined simulator using python to evaluate the performance of our system. The simulator follows the same settings in Table 4.3. The total number of subchannels is still 10, each with 1 subcarrier for identifier transmission. Constant defer mechanism of Identifier Learning is adopted here, where P_{defer} is set to 0.5. We compare AT-Learning with multichannel Slotted-ALOHA, with different number of stations ranging from 10 to 50. In order to focus on the performance on the channel utilization, we assume that the network is saturated and the packet reception failure is only caused by the collisions.

Convergence Time

Figure 4.9 presents the average number of convergence steps before reaching an efficient allocation ("steady state"). We see that convergence time increases as the coordination signal space C increases. This is quite reasonable and consistent with the analysis in Sect. 4.2.2.3. Larger coordination signal space requests more steps for convergence, since as the number of coordination signals increases, stations may have to await a long time for the appearance of a certain coordination signal to transmit. Thus convergence will not be reached until all the stations have transmitted according to all the coordination signals. It is interesting that it takes the longest time to converge when $M = S$, which is also discussed in [10]. Despite $M = S$, the convergence steps increase as the number of stations increases. This is mainly because more stations have relatively high probabilities to introduce collisions. Thus they have to adapt strategy tables more frequently, resulting in a longer convergence time. However, the convergence time is around 10^2 even with 10 correlation signals, which is acceptable for the initialization of a network.

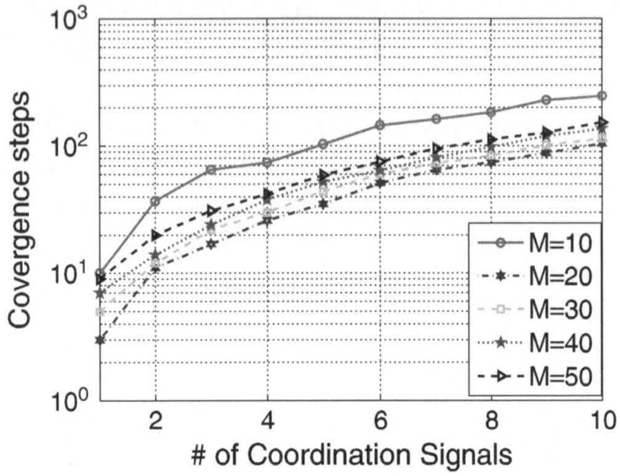

Fig. 4.9 Number of steps
that stations can reach an
efficient allocation

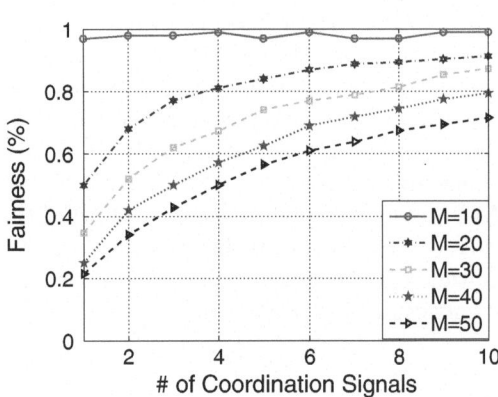

Fig. 4.10 Fairness among
stations under different
number of coordination
signals

Fairness

Fairness is also another important metric to evaluate the feasibility of our scheme.
Recalling in Sect. 4.2.2.3, we deduce that if we choose a relatively large C, $J(X)$ is
close to 1 when the number of clients is close to infinity, which theoretically proves
that our scheme is fair enough. Since in practice we are not allowed to choose a
large value of C as the number of subcarriers is limited, we evaluate fairness using a
small value of C to see how AT-Learning will perform in real-world environments.
As shown in Fig. 4.10, we calculate the average probability of all the stations to
access the channel. When $M = S$, this probability approximates 1, which means
each client in every slot has equal chance to transmit on one channel. As the number
of stations increases, fairness decreases, since there will not be enough channels
for each station to transmit in every time slot. This reduction can be compensated
by increasing the signal space of C. As C exceeds certain threshold, e.g., $C = 5$,
fairness is above 65 % even with 40 stations. This is mainly because with more
coordination signals, stations can balance their strategies with each other. Taking
convergence time and fairness together into consideration, we observe a trade-off to
design a coordination signal space C. C cannot be too large since convergence time

will be too long. C also cannot be too small to ensure fairness. Different systems require different C. For a network with 10 subchannels and 10–30 stations, we consider $C = 6$ according to the simulation results.

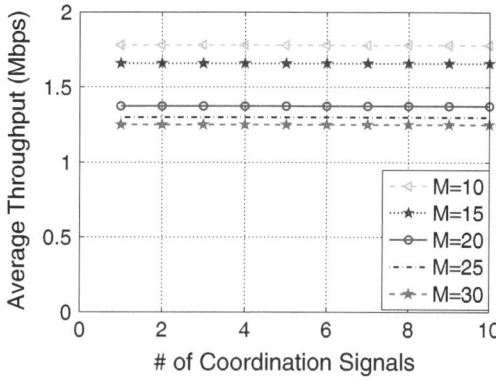

Fig. 4.11 Performance of multichannel Slotted-ALOHA with varying signal space C and different number of stations

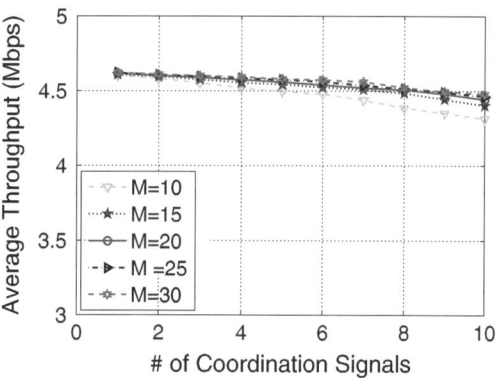

Fig. 4.12 Performance of AT-Learning with varying signal space C and different number of stations

Average Throughput

In this step, we evaluate the average throughput of an AT-Learning system comparing with multichannel Slotted-ALOHA under different number of stations ranging from 10 to 30, which is a typical size of a contention domain. When stations use multichannel Slotted-ALOHA as their access scheme, no carrier sense is performed, they randomly choose subchannels to transmit without learning or adaptation. The reason why we do not choose multichannel Slotted-CSMA is that AT-Learning does not perform carrier sense to check whether the subchannel is busy or not before transmission, it merely relies on a strategy table to determine whether to access or not. And monitoring is only performed when stations do not transmit in a certain time slot. Therefore, it is more like ALOHA to some extent.

Fig. 4.13 The impact of defer probability P_{defer} on the performance of AT-Learning ($M = 15$, $S = 10$)

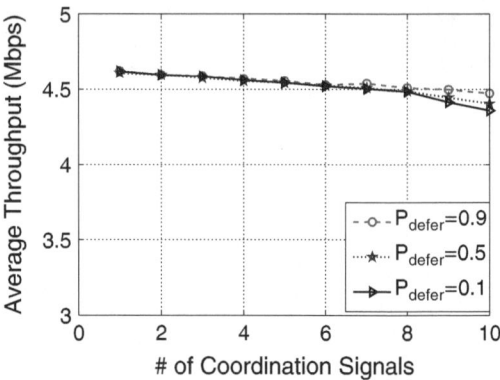

Figure 4.11 shows the average throughput of clients using multichannel Slotted-ALOHA, with different number of stations. Not surprisingly, the throughput is very poor, especially when $M = 30$, which only achieves less than 1.2 Mbps per station comparing with a data transmission rate of 6 Mbps. This is because stations only depend on randomness to avoid transmission collision on each subchannel. Thus collisions happen frequently when the number of stations grows. Figure 4.12 depicts the average throughput using AT-Learning after reaching a steady state. With a data transmission rate of 6 Mbps, the throughput of AT-Learning can achieve about 4.5 Mbps with different number of stations. This verifies that AT-Learning can ensure network performance even under high traffic load. We also study the impact of defer probability P_{defer} on the performance of AT-Learning. From Fig. 4.13 we can see that the value of P_{defer} has little influence on the average throughput when C is smaller than a threshold, e.g., $C = 7$. When C is relatively large, e.g., $C = 10$, it is better to set P_{defer} to a larger value to avoid collision. We calculate the performance gain of AT-Learning over a multichannel Slotted-ALOHA. The gain is up to 300 %. For different number of stations, AT-Learning all performs higher than 4.5 Mbps per station, which is very desirable compared with data transmission rate of 6 Mbps. These results demonstrate that our scheme is efficient to allocate channels for multiple stations.

4.3 Attachment Sense for Hidden and Exposed Terminals

The hidden and exposed terminal problems are two well-known problems in WLANs, which significantly degrade the network performance. As shown by Gollakota and Katabi [13], the hidden terminal problem introduces severe packet loss due to collisions for 10 % of the sender–receiver pairs. Furthermore, in [14], the author shows that the exposed terminal problem can waste useful concurrent transmission opportunities. Extensive research has been carried out to solve these two problems. For example, full duplex [15] allows a receiver to send a busy tune when receiving a data packet. This scheme mitigates the hidden terminal problem, but the exposed node still exists. CMAP [14] deduces the exposed node and excludes a collided transmission by consulting a "Conflict Map," but the hidden terminal problem becomes even more acute. Carrier Sense Multiple Access with Collision Avoidance (CSMA/CA) designs a handshake mechanism called RTS/CTS [16] to mitigate both the hidden and the exposed terminal problems. However, RTS/CTS induces a rather high cost and introduces other problems like false blocking. Therefore, RTS/CTS is disabled by default in WLANs.

When trying to solve both the hidden and the exposed terminal problems, a trade-off arises between collisions (hidden nodes) and unused capacities (exposed nodes). Carrier Sense (CS) is the best effort to resolve this trade-off, but the information obtained (whether the channel is busy or not) is too coarse. We argue that accurate Channel Usage Information (CUI, which nodes are on transmissions or idle nearby) is required to resolve this trade-off. More specifically, PHY layer techniques should be utilized to provide more information about CUI. Then a MAC layer protocol can make the right channel access decision in the presence of the hidden and exposed nodes (Fig. 4.14).

Another emerging technique for wireless transceivers is a full duplex paradigm. This encourages us to propose a cross-layer design, FAST (Full-duplex Attachment System), to solve both the hidden and exposed terminal problems. FAST contains a PHY layer protocol, *Attachment Coding*, which applies attachment transmission to a full duplex paradigm in OFDM-based WLANs, and a MAC layer protocol, *Attachment Sense*, which utilizes the information provided by a PHY layer to make access decisions. Specifically, full duplex attachment transmission provides accurate CUI in real time by letting transmitting nodes modulate their IDs into attachments. Accordingly, *Attachment Sense* instructs nodes to identify the hidden and exposed nodes through online CUI, and thus helps them make the right access decisions fast and accurate.

4.3.1   Attachment Sense Overview

The key insight to solve the hidden and the exposed terminal problems both at once stems from the phenomenon that whether a transmission is successful or not depends only on the channel condition near the receiver side. Therefore, we need a receiver or a victim (a node who is being affected by other transmissions) to claim that they are currently busy within this neighborhood. With the information that who is receiving or being affected nearby, a sender is capable of deferring the transmission to them (*hidden node*). Meanwhile, since a sender does not need to worry about other current senders nearby, it can also conduct concurrent transmissions when

Fig. 4.14 An illustration of
the trade-off when using
Carrier Sense

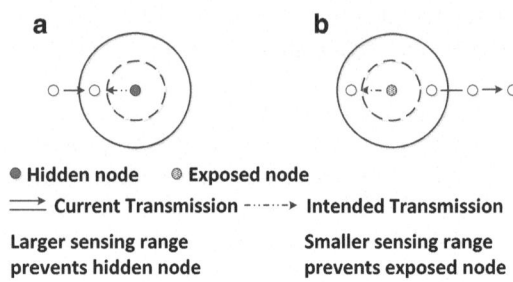

there is no receiver or victim presences (*exposed node*). With this transmission status (CUI) in hand, both the hidden and the exposed terminal problems can be solved.

Inspired by the above observation, we propose an Attachment Sense, which utilizes full duplex attachment transmission to fulfill the above requirements. Specifically, a sender, a receiver, and a victim modulate their identities into attachment and transmit them into the air when they are on transmissions or being affected. These attachments serve as a declaration of current "unavailable" nodes. It is noted that a sender is also required to transmit an attachment along with its data packets. This is to avoid performance degradation by a busy sender (a sender who is transmitting now is also the receiver of other senders). The design principle of attachment sense is simple and efficient, but there remain several implementation challenges. First, an attachment format should be designed efficiently due to the limited bandwidth of each subcarrier. Second, how to make an access decision to resolve the trade-off between the hidden and exposed nodes remains a concern. Last, when utilizing exposed nodes for concurrent transmissions, we should carefully cope with ACK collision with other data transmissions to increase the PRR (Packet Reception Rate).

4.3.2   Attachment Format

The format of attachment should follow several principles. First, different nodes should have exclusive subcarriers for their attachments to avoid confusion. However, since the number of subcarriers is limited, it is not easy to allocate different subcarriers to different nodes in a decentralized manner. Second, it is impossible to modulate the whole identity (MAC address) into attachments due to high bandwidth cost. To address these problems, Attachment Sense has a specialized hash format, which contains the hash value of the corresponding node's ID. Specifically, the whole subcarriers are split into a sender, a receiver, and a victim band. In each band, a membership vector of n subcarriers is used to represent a node identity. This hash format guarantees that an attachment is to be modulated into only one OFDM symbol (e.g., 256-point FFT). When a node transmits its attachment, its MAC address is hashed into a value between 0 to $(n-1)$. Then the corresponding subcarrier in a sender, a receiver, or a victim band will carry a "1" bit. Each

Fig. 4.15 Overview of Attachment Sense. The distributed stations "attach" their identities on their own transmissions to declare the CUI

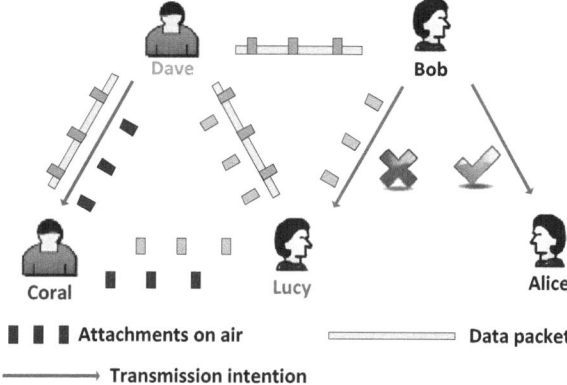

node only needs to acquire the information of the channel usage within one-hop neighborhood (e.g., a degree of 15 in a sparse to medium network). With a reasonably sized n (e.g., 50), a hash value collisions should be small enough.

4.3.3 System Architecture

Unlike CSMA that detects carrier waves before transmitting, Attachment Sense simply asks a node to listen to attachments on air. The attachments are generated according to the following rules: (1) The sender transmits data packets and attachments simultaneously; (2) The receiver transmits attachments once it starts to receive data packets; and (3) The victim transmits attachments when it has been affected by other transmissions nearby.

To make a channel access decision, each node maintains two distributed hash lists, Current Transmission List (CTL) and Neighborhood Hash List (NHL). CTL includes the Current Sender Field (CSF), the Current Receiver Field (CRF), and the Current Victim Field (CVF). It is constructed whenever a node has a packet to transmit. After a node detecting attachments on air for one symbol duration, all the hash values contained in attachments will be decoded and filled into CSF, CRF, and CVF, respectively. NHL simply encodes all the one-hop neighbors' IDs. These IDs are also designed as hash values to reduce the overhead of NHL maintenance. We illustrate how to make a channel access decision using Fig. 4.15 through an example. When Bob has a packet to transmit to Alice [who has the hash value of $H(rev)$], he will first listen to attachments on air. After obtaining CTL in hand, Bob will extract the NHL from his routing table and check the following metric:

$$((CRF \cup CVF) \notin NHL) \cap (H(rev) \notin CSF) \qquad (4.5)$$

If the metric returns true, Bob can confirm his transmission and send packets to Alice immediately. Otherwise, if this metric returns false, Bob has to defer his transmission and keeps listening to attachments until the above metric is satisfied.

4.3.4 Points of Discussion

We finish the description of FAST with a few discussion points. For the issues we talk about below, we broadly describe the potential approaches to cope with them. Nevertheless, it leaves exhaustive discussions in further research.

4.3.4.1 Collision Among Comparable-Length Transmission

The first issue is to resolve collisions among different kinds of transmissions. First, ACK may collide with data packets when utilizing exposed nodes for concurrent transmissions within a neighborhood. To avoid ACK collision with data packets, we split a small portion of the subcarriers from the whole channel, which are only used for an ACK transmission. In this case ACK transmission can be separated from data transmission. Second, collisions may also happen when two senders transmit almost simultaneously. To avoid further collision caused by simultaneous transmissions, a backoff counter and a small backoff window are adopted. When a sender notices a collision took place, it will increase the counter by 1, otherwise, the counter is set to 0. Whenever the counter exceeds a certain threshold, say 3, the sender will back off for a few time slots.

4.3.4.2 Compatibility with Full Duplex Paradigm

A second issue to be discussed is whether attachment transmission is compatible with a full duplex paradigm. According to [15], full duplex is achieved by using *balun passive cancelation* at RX to cancel out self-interference from TX. This process will not be affected by Attachment Cancelation since Attachment Cancelation takes advantage of the null header and tailer to cancel out the attachments on air, which is completely independent from self-cancelation. Moreover, Attachment Coding supports full duplex transmission, where each node can double the throughput by sending while receiving. This lies in the fact that an attachment is transmitted independent from data, and thus will not influence normal data transmission.

4.3.4.3 Impact of Hash List Collision

The third issue is to analyze whether a hash value collision will introduce some performance degradation. Since FAST uses hash values to represent nodes' IDs, different nodes may have the same hash value within a neighborhood. In this case, they cannot be distinguished by other nodes. We define that in FAST, whenever a sender detects a hash value collision, it will always defer the current transmission. This conservative manner successfully avoids data packet collisions and reduces the performance loss. To analyze the actual performance loss due to hash collisions, we use a pair-collision for illustration. Pair-collision can be divided into two cases, as shown in Fig. 4.16: collision within the same-hop [e.g., $H(I) = H(G)$ or $H(A) = H(E)$] and collision between different hops [e.g., $H(A) = H(G)$]. In Case 1, a hash collision results in the same action [defer transmission in Case 1(a), or conducts transmission in Case 1(b)], and thus does not introduce a performance loss. In Case 2, since C is not able to distinguish A from G, deferring transmission wastes exposed terminal opportunity. This probability of performance loss can be derived using a geometry representation. We define the entire two-hop area to be unit 1. Given a node C, and a hash collision pair A/G, P_{lost} should satisfy the following conditions:

- A is receiving a data packet in a white area. This probability can be expressed as: $P(A) = \frac{3}{4} \times \frac{1}{3}$;
- G is idle in a green area without intersection with a red area. This probability can be expressed as: $P(G) = \frac{1}{4} \times \frac{2}{3} \times \frac{1}{3}$.

Fig. 4.16 Illustration of hash value collision

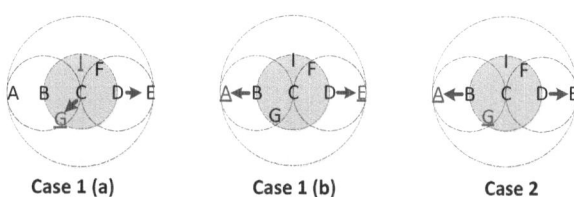

Case 1 (a)   Case 1 (b)   Case 2

Then the probability of missing a concurrent transmission opportunity, $P_{Lost} = P(A) \times P(G) = \frac{1}{72}$, which is relatively small. More number of hash collisions (e.g., triple-collision) can be proven using similar methodology, which has even a smaller probability to introduce a performance loss. Therefore, hash collisions can be harmless.

4.3.5 Performance Evaluation

In this section, we conduct extensive simulations to evaluate the performance of FAST using network simulator NS-3. As illustrated in Fig. 4.17, the simulations are divided into two parts: (1) baseline topology, including hidden nodes, interfering nodes and exposed nodes configuration and (2) practical networks, which is ad-hoc network configuration. The first configuration serves as a baseline to see whether FAST can make the right access decisions in particular scenarios. The second configuration evaluates FAST in practical networks with multiple sender–receiver pairs.

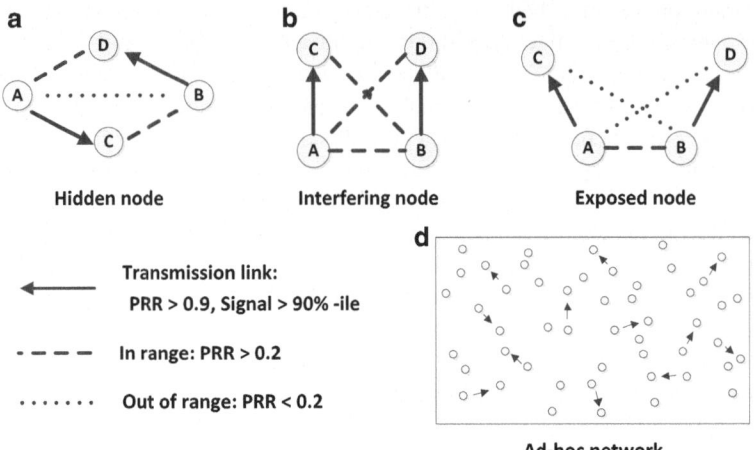

Fig. 4.17 Topologies overview, (**a**), (**b**), and (**c**) baseline topology in Sect. 4.3.5.1, and (**d**) practical networks in Sect. 4.3.5.2

Table 4.3 Experimental configuration parameters

Parameters	Values	Parameters	Values
SIFS	$16\,\mu s$	DIFS	$32\,\mu s$
Symbol time	$16\,\mu s$	Slot time	$9\,\mu s$
CW_{min}	16	CW_{max}	1,024
Packet length	1,460 bytes	Basic data rate	6 Mbps

For the above simulation scenarios, channel bandwidth is 20 Mbps with 256-point FFT OFDM modulation, where 192 and 8 subcarriers are used for data and ACK, respectively. Detailed parameters are shown in Table 4.3, following the specification of 802.11a. Each simulation lasts 50 s. The aggregate throughput is calculated at all the designated receivers. We compare FAST to 802.11 MAC with *CS on* ("ideal case" for an interfering node) and *CS off* ("ideal case" for an exposed node).

4.3.5.1 Baseline Topology

In this part, the performance of FAST is evaluated with two sender–receiver pairs in three basic topologies, as shown in Fig. 4.17a–c. We select the above three configurations from a general 50-node topology with random distribution and degree of 12 (average number of neighbors) in Fig. 4.17d. Each configuration is repeated 50 times, with different sender–receiver pairs each time. The selection principles are shown in Fig. 4.17, which are trained in advance and recorded for each link.

Fig. 4.18 Aggregate throughput for hidden node configuration

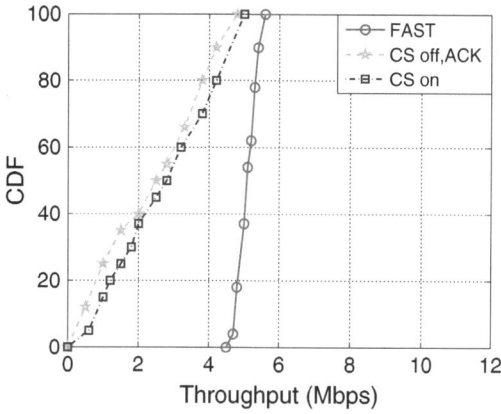

Fig. 4.19 Aggregate throughput for interfering node configuration

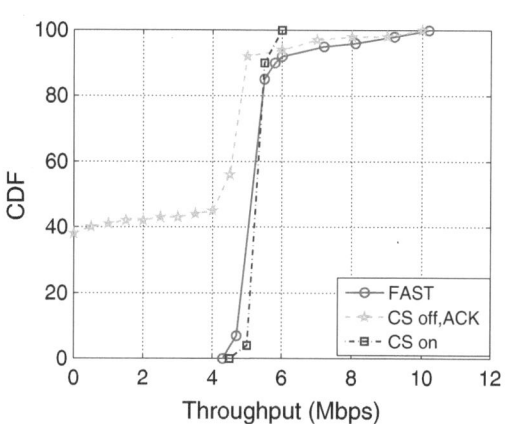

Hidden Node

Figure 4.18 shows the performance of *CS on*, *CS off*, and FAST in a hidden node configuration. Ideally, there should only be one transmission at a time. *CS on* is unable to identify whether other nodes are receiving data packets within a neighborhood. Meanwhile, *CS off* merely transmits into air no matter there is any other receiver within a transmission range. Therefore, nodes in *CS on* and *CS off* collide frequently, resulting in a median throughout of less than 3 Mbps. Fortunately, a backoff strategy mitigates the performance degradation from collisions, and there are about 20 % nodes achieving a throughput of 4 Mbps. On the other hand, FAST instructs nodes to identify current receivers nearby (node C and D) through their attachments, and thus prevents hidden nodes (node A and B) to transmit concurrently. In this case, nodes transmit one after another, and can achieve a throughput of 5.2 Mbps, which approximates the ideal performance for a hidden node configuration.

Interfering Node

Figure 4.19 shows the performance of *CS on*, *CS off*, and FAST in a exposed node configuration. *CS off* achieves zero throughput for over 40 % of the link pairs, where concurrent transmissions are completely corrupted by each other. For the rest 60 % of the link pairs, interfering simultaneous transmissions give*CS off* very poor

Fig. 4.20 Aggregate throughput for exposed node configuration

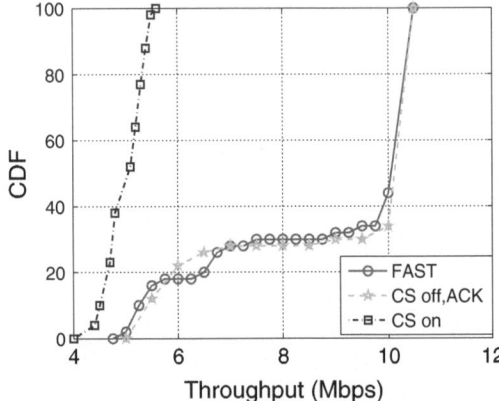

performance. On the contrary, FAST correctly figures out interfering transmissions through attachments from victims, and let senders take turns to transmit. In this case, FAST can achieve almost a single-link throughput as *CS on*. There are 8 % of the link pairs achieve double-link throughput for *CS off* (on top of the figure), indicating that they are actually exposed nodes. For these link pairs, FAST quickly traces the curve of *CS off* and achieves a similar performance.

Exposed Node

Figure 4.20 shows the performance of *CS on*, *CS off*, and FAST in a exposed node configuration. From the dash-dot line with squares we can see that *CS on* prevents exposed nodes from transmitting concurrently. Thus most of the link pairs only achieve single-link throughput of 5 Mbps. With *CS off* and ACK disabled, 27 % of the link pairs achieve little more than single-link throughput, revealing that they are not actually exposed nodes. For the rest 73 % of the link pairs, *CS off* leverages exposed nodes to achieve double-link throughput up to 10.5 Mbps. FAST traces well the curve of *CS off*, indicating that through accurate CUI, Attachment Sense fully utilizes exposed nodes. It is noted that FAST has little performance reduction comparing with *CS off* (about 0.2 Mbps), for the reason that ACK is disabled in *CS off*, but FAST still has ACK overhead to the overall throughput.

4.3.5.2 Practical Networks

In this part, we quantify the performance of FAST in ad hoc networks [17, 18], as illustrated in Fig. 4.17d. In ad hoc networks, hidden and exposed nodes significantly degrade the network performance, especially with high node density and heavy traffic load [19]. We choose 6, 8, 10, and 12 number of concurrent senders as four configurations. Each configuration runs 50 times, and each time with different senders transmitting simultaneously with no more constraints.

We calculate the per-sender throughput for FAST, *CS on* and *CS off* in each configuration. By preventing hidden nodes from collisions and exploiting exposed nodes for concurrent transmissions, FAST improves per-sender throughput over *CS on* by between 180 % ($N = 6$) and 200 % ($N = 8$), and over *CS off* by between 200 % ($N = 6$) and 220 % ($N = 8$). When the number of concurrent transmissions increases, nodes may transmit simultaneously and introduce unavoidable collisions, resulting in small performance degradation for FAST. However, it still improves the performance over *CS* by over 200 % (N = 12). Therefore, FAST is promising in dense networks and can achieve much better throughput over 802.11 CSMA (Fig. 4.21).

Fig. 4.21 Per-sender throughput in ad hoc networks with different # of concurrent transmissions

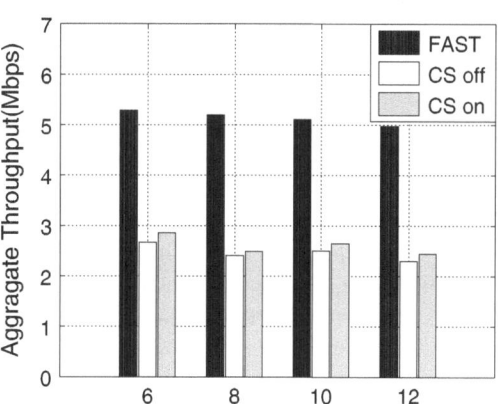

4.4 Performance Evaluation for Variable Bit-Rates

802.11a provides different bit-rates from 6 to 54 Mbps. In practice, higher bit-rates rather than 6 Mbps are utilized to increase throughput. Therefore, whether FAST can still decode attachments and leverage exposed terminals in higher bit-rates remains concern. We repeat the simulation of static ad hoc network configuration in the previous simulations with the same setting, except transmission rate. Each time we choose three concurrent transmission pairs from all possible nodes, and let them transmit with different bit-rates from 6 to 54 Mbps.

Fig. 4.22 Throughput gain
of FAST over CSMA and
CMAP in static ad hoc
networks under different data
rates, N = 8

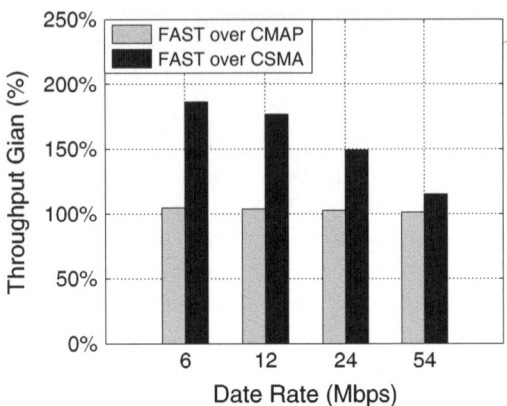

We compute the aggregate throughput for FAST, CSMA, and CMAP and depict
the performance gain of FAST over the other approaches in Fig. 4.22. With bit-
rate of 6 Mbps, FAST achieves up to 186 % performance gain over CSMA, and
a little performance gain over CMAP. As the data rate increases, the performance
gain decreases. That is because with higher data rate, the decodability of attachment
drops. Therefore, senders may not be able to get the accurate information of the
ongoing transmission. Concurrency cannot be fully utilized, and even collision may
happen. However, FAST can still achieve 115 % throughput gain over CSMA, which
is similar as CMAP. FAST has little throughput improvement over CMAP in static
ad hoc networks, as we discussed above, even with higher data rates. This verifies
that CMAP needs to consume extra resource for "conflict map" construction, while
FAST can obtain CUI cost-efficiently. Another point is that as transmission rate
increases, reception range and interference range diverge significantly. Therefore,
transmission of a sender may interfere the nodes out of its one-hop neighborhood.
In this case, concurrent transmission by exposed terminal is better not to be utilized,
since it may introduce more collisions. To deal with this problem, we propose a
simple solution. When data rate exceeds certain threshold (e.g., 36 Mbps), nodes
will "turn off" the exposed terminal function. It means even a node identifies itself
as an exposed terminal, concurrent transmission will not be leveraged. This solution
may constrain the usage of exposed terminal. We will leave it for our future work.

References

1. Y. Bejerano, H.-G. Choi, S.-J. Han, and T. Nandagopal, "Performance tuning of infrastructure-
   mode wireless lans," in *Modeling and Optimization in Mobile, Ad Hoc and Wireless Networks
   (WiOpt), 2010 Proceedings of the 8th International Symposium on*, pp. 60–69, IEEE, 2010.
2. S. Katti, S. Gollakota, and D. Katabi, "Embracing wireless interference: Analog network
   coding," in *ACM SIGCOMM Computer Communication Review*, vol. 37, pp. 397–408, 2007.

3. Y. C. Du Ho Kang, D. Kim, and S. Choi, "Qos-aware load indicators for intelligent cell selection," *IEEE APWCS*, 2010.
4. I. . W. Group *et al.*, *IEEE 802.11n-2009: Enhancements for Higher Throughput*, 2009.
5. G. Zhou, C. Huang, T. Yan, T. He, J. Stankovic, and T. Abdelzaher, "Mmsn: Multi-frequency media access control for wireless sensor networks," in *IEEE Infocom*, pp. 1–13, 2006.
6. P. Mähönen and M. Petrova, "Minority game for cognitive radios: Cooperating without cooperation," *Physical Communication*, vol. 1, no. 2, pp. 94–102, 2008.
7. L. Gao and X. Wang, "A game approach for multi-channel allocation in multi-hop wireless networks," in *Proceedings of the 9th ACM international symposium on Mobile ad hoc networking and computing*, pp. 303–312, ACM, 2008.
8. J. Park and M. Van Der Schaar, "Medium access control protocols with memory," *IEEE/ACM Transactions on Networking (TON)*, vol. 18, no. 6, pp. 1921–1934, 2010.
9. L. Cigler and B. Faltings, "Reaching correlated equilibria through multi-agent learning," in *The 10th International Conference on Autonomous Agents and Multiagent Systems*, vol. 2, pp. 509–516, 2011.
10. R. Aumann, "Subjectivity and correlation in randomized strategies," *Levine's Working Paper Archive*, 2010.
11. R. Jain, D. Chiu, and W. Hawe, "A quantitative measure of fairness and discrimination for resource allocation in shared computer systems," *DEC research report TR-301*, 1984.
12. S. Sen, R. Choudhury, and S. Nelakuditi, "Listen (on the frequency domain) before you talk," in *ACM SIGCOMM Workshop on Hot Topics in Networks*, p. 16, 2010.
13. S. Gollakota and D. Katabi, "Zigzag decoding: combating hidden terminals in wireless networks," in *ACM SIGCOMM Computer Communication Review*, vol. 38, pp. 159–170, 2008.
14. M. Vutukuru, K. Jamieson, and H. Balakrishnan, "Harnessing exposed terminals in wireless networks," in *Proceedings of the 5th USENIX Symposium on Networked Systems Design and Implementation*, pp. 59–72, 2008.
15. M. Jain, J. Choi, T. Kim, D. Bharadia, S. Seth, K. Srinivasan, P. Levis, S. Katti, and P. Sinha, "Practical, real-time, full duplex wireless," in *ACM MobiCom*, pp. 301–312, 2011.
16. I. . W. Group *et al.*, *IEEE 802.11-2007: Wireless LAN Medium Access Control (MAC) and Physical Layer (PHY) Specifications*, 2007.
17. W. Yu, J. Cao, X. Zhou, and X. Wang, "A high-throughput mac protocol for wireless ad hoc networks," in *IEEE international conference on Pervasive Computing and Communications Workshops (PERCOMW)*, pp. 405–pp, 2006.
18. M. Dianati, X. Ling, K. Naik, and X. Shen, "Performance analysis of the node cooperative arq scheme for wireless ad-hoc networks," in *IEEE GLOBECOM*, vol. 5, pp. 5–pp, 2005.
19. S. Zhao, L. Fu, X. Wang, and Q. Zhang, "Fundamental relationship between nodedensity and delay in wireless ad hoc networks with unreliable links," in *ACM MobiCom*, pp. 337–348, 2011.

Chapter 5
Conclusion and Future Work

5.1 Conclusion

In this book, we present a novel attachment transmission paradigm, which enables data packets transmitted along with control messages, without degrading the throughput of the original data traffic. We theoretically analyze the feasibility and reliability of attachment transmission, and for verification also conduct experiments via GNU Radio testbed. As stated in Chap. 3, experimental results show that attachment transmission does have little impact on the original data transmission. However, with as many as six concurrent attachment transmissions, the performance loss is all under 10^{-2}. As for the reliability, the miss detection rate of attachment transmission can be controlled to within 1%, resulting in a detection accuracy of more than 99%.

To illustrate the effectiveness of attachment transmission, we propose several new cross-layer designs to solve the classic problems in wireless networks, including multiple access problem, multichannel allocation problem, and the hidden and exposed terminal problems. Taking advantage of attachment transmission, harmless attachment can improve the efficiency of infrastructure-based WLANs by up to 200% compared with the existing 802.11 family protocols. In a distributed OFDMA-based networks, attachment learning can achieve a fair and efficient channel allocation and improve the throughput by up to 300% over Slotted ALOHA. Last but not least, by preventing hidden terminal and harnessing exposed terminals, attachment sense improves the average throughput by up to 200% over CSMA in practical ad hoc networks.

L. Wang et al., *Attachment Transmission in Wireless Networks*,
SpringerBriefs in Computer Science, DOI 10.1007/978-3-319-04909-0_5,
© The Author(s) 2014

5.2 More Opportunities in Attachment Transmission

Besides the above-mentioned problems in wireless networks, we believe that attachment transmission can benefit more communication systems. In this section, we broadly offer some potential topics that may leverage attachment transmission to improve the performance, but in general we leave an exhaustive discussion of them as future work.

5.2.1 Attachment Transmission for QoS Control

Multimedia transmission in wireless networks is receiving a lot of attention these days, which has heterogeneous data traffic and different transmission require-ments [1,2]. For example, Constant Bit Rate (CBR) traffic prefers reservation-based MAC, which can ensure determined performance. Variable Bit Rate (VBR) traffic prefers contention-based MAC, which can reduce transmission delay [3]. Current single MAC protocols cannot meet the above requirements and utilize the diversity of different traffic types. This trend calls for a hybrid MAC protocol, which can support and utilize different types of traffic for transmission. We split the whole channel into different class of subchannels, each class for one type of transmission. Nodes adopt their preferred MAC protocols on different class of subchannels. However, to allocate different subchannels to different types of traffic dynamically is not trivial in distributed networks. Current PHY layer techniques cannot provide so much information for the MAC layer to do this. Therefore, we can utilize attachment transmission for nodes to declare their transmission requirement in attachments. After obtaining the whole requirements within a neighborhood, nodes can compute a distributed channel allocation strategy and utilize their preferred MAC for transmission.

5.2.2 Attachment Transmission for Coordination in CRNs

During the last two decades, radio spectrum resource is shown to be significantly underutilized with fixed spectrum assignment policy. Cognitive radio emerges as a promising solution, which allows unlicensed users to opportunistically access the spectrum not used by the licensed users [4]. To ensure that unlicensed users can identify vacate spectrum fast and accurately without interfering licensed users, cooperative sensing is explored to improve the sensing performance by leveraging spatial diversity. However, the design of CR networks imposes unique challenges due to the high fluctuation in the vacant spectrum and the opportunistic access among CR users. The first challenge is to accurately identify the available spectrum in real time through spectrum sensing, while vacate the spectrum once the PU is detected. This sensing accuracy is compromised with many factors, such as multi-path fading and shadowing [5]. Recently, cooperative spectrum sensing has shown its superiority to improve the sensing accuracy by exploiting spatial diversity. After exchanging sensing information among spatially located CR users, each of them makes a combined decision, which can be more accurate than individual ones. However, cooperation overhead increases dramatically and comprises the sensing performance, especially in distributed networks. The second challenge is to share the available spectrum among different CR users once the sensing decisions have been made. As the available spectrum and node density increases, coordination overhead and transmission delay raise up accordingly, resulting in a significant performance degradation. These challenges necessitate efficient designs that can simultaneously address extensive communication problems in CR networks. To reduce the control overhead in cognitive radio networks (CRNs), attachment transmission can be utilized to carry cost-effective sensing decision and channel allocation information. One penitential challenge is the synchronization issue in distributed CRNs. As it is known that symbol level synchronization is a requirement for OFDM modulation, we leave it as future work to study an attachment transmission paradigm that is less synchronization sensitive for distributed CRNs.

References

1. C. Zhu and M. S. Corson, "Qos routing for mobile ad hoc networks," in *INFOCOM 2002. Twenty-First Annual Joint Conference of the IEEE Computer and Communications Societies. Proceedings. IEEE*, vol. 2, pp. 958–967, IEEE, 2002.
2. Y. Wu and D. H. Tsang, "Distributed power allocation algorithm for spectrum sharing cognitive radio networks with qos guarantee," in *INFOCOM 2009, IEEE*, pp. 981–989, IEEE, 2009.
3. R. Zhang, R. Ruby, J. Pan, L. Cai, and X. Shen, "A hybrid reservation/contention-based mac for video streaming over wireless networks," *Selected Areas in Communications, IEEE Journal on*, vol. 28, no. 3, pp. 389–398, 2010.
4. C. Gao, Y. Shi, Y. Hou, H. Sherali, and H. Zhou, "Multicast communications in multi-hop cognitive radio networks," *IEEE Journal on Selected Areas in Communications*, vol. 29, pp. 784–793, 2011.
5. I. F. Akyildiz, B. F. Lo, and R. Balakrishnan, "Cooperative spectrum sensing in cognitive radio networks: A survey," *Physical Communication*, vol. 4, no. 1, pp. 40–62, 2011.

Index

Attachment
  Attachment Coding, 47
  Attachment Learning, 35
  Attachment Sense, 47
  Attachment Transmission, 17
  Harmless Attachment, 29

CAS, 9
  GCAS, 10
  ICAS, 10
  SCAS, 9
Channel Sense
  CSMA, 2
  CSMA/CA, 2
Classic Problem
  Exposed Terminal Problem, 48

Hidden Terminal Problem, 47
Multichannel Allocation Problem, 36

Equilibrium
  Correlated Equilibrium, 35
  Nash Equilibrium, 35
Exposed Terminal, 1

Hidden Terminal, 1

Modulation
  OFDMa, 9
  OFDN, 7

L. Wang et al., *Attachment Transmission in Wireless Networks*,
SpringerBriefs in Computer Science, DOI 10.1007/978-3-319-04909-0,
© The Author(s) 2014